Quantitative Reasoning

Is college worth the cost? Should I worry about arsenic in my rice? Can we recycle pollution? Real questions of personal finance, public health, and social policy require sober, data-driven analyses. This unique text provides students with the tools of quantitative reasoning to answer such questions. The text models how to clarify the question, recognize and avoid bias, isolate relevant factors, gather data, and construct numerical analyses for interpretation. Themes and techniques are repeated across chapters, with a progression in mathematical sophistication over the course of the book, which helps the student get comfortable with the process of thinking in numbers. This textbook includes references to source materials and suggested further reading, making it user-friendly for motivated undergraduate students. The many detailed problems and worked solutions in the text and extensive appendices help the reader learn mathematical areas such as algebra, functions, graphs, and probability. End-of-chapter problem material provides practice for students, and suggested projects are provided with each chapter. A solutions manual is available online for instructors.

Eric Zaslow is a Professor of Mathematics at Northwestern University. He has a Ph.D. from Harvard University in mathematical physics.

Quantitative Reasoning

Thinking in Numbers

ERIC ZASLOW
Northwestern University, Illinois

CAMBRIDGE
UNIVERSITY PRESS

CAMBRIDGE
UNIVERSITY PRESS

University Printing House, Cambridge CB2 8BS, United Kingdom

One Liberty Plaza, 20th Floor, New York, NY 10006, USA

477 Williamstown Road, Port Melbourne, VIC 3207, Australia

314–321, 3rd Floor, Plot 3, Splendor Forum, Jasola District Centre,
New Delhi – 110025, India

79 Anson Road, #06–04/06, Singapore 079906

Cambridge University Press is part of the University of Cambridge.

It furthers the University's mission by disseminating knowledge in the pursuit of
education, learning, and research at the highest international levels of excellence.

www.cambridge.org
Information on this title: www.cambridge.org/9781108419413
DOI: 10.1017/9781108297844

© Eric Zaslow 2020

First published 2020

Printed in the United Kingdom by TJ International Ltd. Padstow Cornwall

A catalogue record for this publication is available from the British Library.

ISBN 978-1-108-41941-3 Hardback
ISBN 978-1-108-41090-8 Paperback

Contents

Preface

This book serves a course designed to teach first-year college students to try to answer topical, real-life questions using numerical analysis, or *quantitative reasoning*. The goal is to develop a student's ability to make carefully reasoned, quantitative arguments utilizing basic mathematical skills. Quantitative reasoning is universally important in all scholarship, and to drive this point home, we have chosen motivating questions from a variety of real-world concerns and a host of academic disciplines. These arguments form the core of the book and are buttressed by skills developed in the appendices. We hope that the topics we explore along the way are both stimulating and fun for the reader.

In many cases, a reliable answer to the questions we pose would require a major effort by an academic researcher. Our aim is not perfection[1] but to familiarize ourselves with the methods and practices of structured, numerical arguments. The pieces of these arguments are usually "simple" aspects of mathematics or statistics. We assume a familiarity but not a facility with these skills. The appendices offer brief reviews of the essential concepts and hints to their use, but the real focus is the application of these skills to answering the questions posed. The pedagogy comes through the extensive, worked examples that compose the chapters. Much of the discussion concerns the tasks of creating a well-posed question, gathering reliable research, constructing a model, and recognizing its underlying assumptions and limitations.

Many themes and techniques are repeated in multiple chapters, although there is a general progression in mathematical sophistication over the course of the book as the topics become more scientific. We view scientific exploration as no different from any field of inquiry and make no attempt to separate it out. The book adopts a narrative style, an intentional contrast from the math-first approach of many curricula.

Answering a variety of questions in essentially the same way should help the student get comfortable with the process of "thinking in numbers." While individual mathematical skills may be practiced and mastered, it is the *coherent assembly* of

---

[1] Indeed, our use of economics and social science is pretty shoddy by academic standards. Real-world problems have too many variables for a tidy and rigorous treatment aimed at a novice. Our hope in addressing topical albeit messy questions is that even a somewhat simplified analysis of an actual question will be illuminating and educational.

these skills that quantitative reasoning demands. Professors often expect this ability in their students, and aspects of quantitative reasoning are treated in economics, statistics, and mathematics courses. But few university or high school curricula have courses dedicated to this purpose: this book is a *one-stop shop* for the reader to review or learn basic mathematical skills while learning how to apply them to multistep arguments in a quantitative way. The appendices serve to remind and instruct the reader in the necessary mathematical areas, such as algebra, functions, graphs, probability, and statistics. Readers should be careful to read mathematics slowly. Glossing over the numbers in favor of the text will render the book useless.

Teacher's Manual

The book is written for a quarter or semester course.

Syllabus. For a ten-week quarter course, an instructor could first review the introduction, then proceed by covering one chapter per week, together with a corresponding appendix section, usually with the same number (e.g., Week 1: Chapter 1, Appendix 1). Each of Appendices 7 and 8 can be spread out over two weeks, and Appendix 9 questions may be assigned at any time. The match-up of material won't be perfect, but overlap and review are beneficial to the students. Most of the appendices will be review for most of the students, anyway.

For a semester course, the instructor can spend a full week on the introduction, then do one chapter each week along the lines above, but with two weeks each on Chapters 9 and 10.

Modularity. An important feature of the design is that the chapters are independent enough so that an instructor can skip some along the way, if time is running short. This has the benefit of allowing an instructor to delve into any aspect of a chapter without worrying about paying service to a rigid curriculum. If I need to cull material, I like to poll my class to see which chapters interest them most, then adjust the schedule accordingly.

Homework. Homework should be assigned, collected, and graded each week: students cannot learn quantitative skills without practice and assessment. A typical homework assignment could be all of (or, more realistically, most of) the exercises from a chapter and a corresponding appendix section, usually with the same number.

It is recommended that the instructor also ask the students to "pre-read" the week's chapter for general concepts (not for specific quantitative skills), then begin the week with a short quiz, 3–5 minutes, testing only rough comprehension: anyone who has done the reading should do well on the quiz.

Projects. Students should be assigned projects over the course of the class. Project suggestions are listed at the end of each chapter, though students may have their own ideas. A project report will look roughly like one of the chapters in the text, though not as pedagogical. In a ten-week course, students might be expected to complete two projects (typically as a group), while a semester-long course might assign three. In either case, the first project should be on the lighter side, as the students will be new to both qualitative and quantitative aspects of reasoning early in the term. I sometimes ask for just an outline for the first project. Instructors and/or teaching assistants should meet with individuals or groups to guide them in beginning their projects, then several times along the way to help ensure they reach some form of conclusion. Try to allot time for in-class presentations of final projects – the students really enjoy it.

Teachers may direct students to the following (as they won't be reading this far into a preface).

To the Student
Here's all you need to know:

- READ WITH A PEN AND PAPER! You can't read anything with numbers unless you're constantly jotting them down, verifying that you agree with what is written, and deriving results for yourself.
- Go slowly! Read a bit; think a lot. The writing is dense. Points are made, but not repeated. There is no "busy reading." The price tag for this is that you need to consider what is being discussed until it is clear to you.
- The math is in the back. If you're reading something and the math of it is getting too tricky, go the relevant appendix and bone up on your skills. A footnote at the start of each chapter directs you to the most relevant appendix.
- Answer the questions embedded in the text; do the exercises at the end of chapters. The course has homework. Do it. You won't learn anything unless you actually work through examples. The conclusions of the chapters – yes, the sky *is* dark at night! – are meaningless if you haven't developed the ability to explain how any of them are formed.
- "Leave it on the field." One chapter might be difficult. Keep going, and don't let it get to you. Lots of the material will show up again in a later chapter. Over time, you will gain a facility for the techniques involved in the course. ("Leave it on the field" can also mean "try hard in the moment." I assume you will!)

The words above are specific to this course, but as we emphasize the universal nature of reasoning, we should also note that there are universals, best practices, for attending *any* college course:

- Go to class. That is where instructors communicate their deep understanding of the subject to you. Take notes!
- Go to office hours. First, try hard on your own. Then, unless you are sailing through easily, you are likely to need help, and direct contact (not e-mail) with your instructors and teaching assistants is the best way to get it.
- Self-assess. Be frank with yourself about what you do and don't understand (your graded assignments are a good guide). Try to correct what you don't, even if it

means creating extra work for yourself beyond the assignments. Most instructors expect you to do this on your own but may not say so explicitly.

- Talk to classmates and fellow students, but know that you have only understood something when you can do it without guidance.
- Stay confident! Struggle is natural. Meaningful accomplishments come through struggle. Do not lose confidence just because the going gets tough.

Finally, a comment to the reader: the book will try to speak to you in a familiar tone, but will inevitably fail. We all come to life from different cultures and backgrounds and with different experiences and perspectives. A writing passage might sound like a "fresh voice" to some but ring an offensive tone to others. Some may bristle at certain subject matters or turns of phrase. I take responsibility for any offense and apologize. At the same time, I ask readers, if this happens, please do not let *my* shortcomings stand in the way of *your* education!

Acknowledgments

It is a pleasure to thank my writing assistants, Aidan Perreault and Emmanuel Rockwell, who worked tirelessly to do research, enrich the text, create exercises and solutions, and make general improvements throughout. I thank the Alumnae Board of Northwestern University for the Curriculum Development Award that helped enable this project to come together. I would like to thank Professor Indira Raman for patiently explaining to me how the eye works. Finally, I am grateful to Lane Fenrich and Mary Finn of the Bridge Program of Northwestern University for supporting the idea of a course in quantitative reasoning.

Introduction

The Case for Quantitative Reasoning

> It's the middle of June and you're running late to meet your dad, so you dash in to the national-brand drug store to see if you can find a last-minute Father's Day gift. A display for a wet-dry razor catches your eye: it's on sale for $44.99. The price is a bit more than you wanted to spend, but it's the right thing at the right time.
>
> Your dad loves it.

Lucky for you, eh? Well, actually, there was a lot more to it.

In fact, something extraordinary just happened. Behind the scenes in this scenario were people in the corporate offices analyzing purchasing habits of millions of customers. They tabulate the effect of pricing on the chance of making a sale, as well as placement of the product on the shelves, type of display, etc. They also look at patterns of shoppers near the holidays to determine which items to promote, and how. What was to you a fortuitous find was to them a carefully planned, successful promotion. While a "mom and pop" store can take chances with their promotions and win some, lose some, a national chain with thousands of stores has to invest large sums of money in preparing a promotion, so it can't afford to go by gut and whimsy. Numerical analysis of consumer data can be both crucial to profits and beneficial to the consumer.

Quantitative reasoning helps in more than just sales. It can address everyday questions of home finances or personal politics. In academics, some fields are *built* for a quantitative approach: economics, statistics, the physical sciences, psychology, social sciences, environmental studies. Other fields, such as history (see Chapter 2) and art, have more surprising connections. Students of every stripe – *people in general* – have much to gain from being able to think with numbers.

Constructing Analytical Models

Understanding human behavior, the environment, and science and technology in the age of Big Data demands quantitative, analytical reasoning. The modern university must shape its curriculum to meet this demand. High school students are trained in mathematics but rarely in the process of assembling mathematical arguments to address real-life questions from different disciplines: personal finance (*Can I afford this house?*), public health (*How does not washing your hands affect your risk of catching cold?*), or social justice (*Does the pattern of employment reveal discrimination?*). Even whimsical questions (*Can I hear a pin drop?*) can lead to serious scholarly inquiry.

In this book, we address this need by exploring several topical questions in depth, the aim being to teach the reader how to approach questions in a reasoned, quantitative way. We don't just respond with our gut. Surely, emotions play a large role in decision-making. We are humans: we act impulsively, react instinctively; we indulge a good feeling and we hold grudges. The role of emotions must be acknowledged and addressed – even incorporated as a decision-making factor – but *after* all this, we must strive for a dispassionate analysis. The analyses we build are constructed after interpreting the questions in a suitable way. We must understand what the question means to us before attempting to build a response. Each chapter explores one question, crafting a quantitative argument through a comfortable narrative, gathering data, and developing an analytical model along the way. Conclusions are drawn afterward.

Here is a rough template for addressing one of our questions.

1 Frame the Question
   a What are your gut biases/opinions? You'll have to separate these from the analysis.
   b Is the question clear? Does it need to be more specific? Revise/interpret as necessary.
   c What will an answer to this question look like?

2 Build the Model
   a What factors will you involve in your analysis, and why? Why have you omitted others?
   b What numerical data will you need to gather? What sources are you using?
   c What assumptions will you make to be able to model the problem?[2]
   d Having isolated relevant factors, how will you construct your model?
   e What are the sources of errors? How large are they? Be as quantitative as possible.
   f Explain your calculation and produce your result.

3 Analyze Results
   a Check your work!
   b With what degree of certainty can you assert your answer? Be as quantitative as possible.
   c Analyze your result. What does it tell you?

---

[2] Most problems will require simplifying assumptions, though some may have an exact answer.

d Reality check: is the result reasonable, or does it hint at errors? (If the latter, go to Step 2f.)

e Write a summary.

Each question is unique, but the broad patterns of analysis are well represented in the above blueprint. Even initially simple questions can be addressed in this way — and when we consider them, complexities emerge.

As a basic example, we may ask, *Which costs less, driving to work or taking the train?* If the grumpy conductor terrifies you (1a), you migh interpret "cost" to include emotional torment — which is okay, if you're up front about it. So maybe you're biased to taking the car, but if you decided to measure cost in *dollars,* your bank account will only know the bottom line. We need to make clear (1b) that we mean, say, economic cost. That settled, we then ask (1c), what could an answer look like? Are we looking for a simple car/train recommendation? Do we want to know the amount saved? Or are we looking for a more refined answer: if you're driving such and such a model, then your cost is whatever. Or maybe a probabilistic response: you are this percent likely to save this much money on the train. Okay, for simplicity, suppose we just want a simple car/train recommendation.

Next (2a), we brainstorm the factors that could contribute: cost of transportation, certainly, but also there is the cost of time. The train may take longer, but perhaps you can use it productively by doing work in your seat. Other factors could be the cost of wear and tear on the vehicle or the

Sidebar to End All Sidebars

Sidebars give you short recipes for doing anything. But real problems are complex, and cookie-cutter approaches may not apply. Examples:

- A word problem: you have two water buckets at a well, an eight-gallon and a thirteen-gallon. You want exactly one gallon. What to do? This problem is tough not because it's hard to isolate the relevant information — it's just a hard problem!

- Police police police police police. You see, the "police police" are those who patrol police. Who polices them? Police police police. So the sentence says that this last group is policed by police. Sometimes the complexity is deep and you simply need to slog through.

- Taxation. You want to analyze the result of a tax hike, but there are too many unknowns and variables and not enough data — and if you make a simplifying assumption, you oversimplify so as to become unhelpful. This happens a lot to us in this course, but if you are a politician or professional analyst, you must ultimately decide.

These bullet points are meant to convince you that you cannot reduce all concepts to bullet points. If they didn't, then maybe I just proved my point!

cost of dry cleaning from the train if you always seem to get coffee and calzones spilled on you. We might also consider societal costs: the environmental impacts of the two choices are different, and this can translate in different costs to society, which, ultimately, citizens like you will have to absorb. And so on. To continue the analysis (2b), we collect the facts and figures of transportation, time, wear and tear. To go further, we assume (2c) that we have identified all the factors or that the other factors are negligible by comparison, and we assume, or argue why, our valuations of the various costs are correct. With these factors and assumptions, we will construct our model (2d) by adding up the costs of the train and comparing them to the costs of driving.

After explaining errors in our cost estimation (2e), we move on to adding costs up (2f) and then (3a) checking our calculations. Suppose we determine that driving is cheaper because we need to be at the office to work and we earn money only for the time punched on the clock, and that factor dominated the transportation and other costs. The certainty (3b) of this conclusion may be limited by the fact that we can't determine the driving time with pinpoint accuracy, and if we are quite sophisticated or have good access to data, we may even be able to estimate with some probability what the range of drive times is likely to be and how these might affect (or might not) our recommendation. Now we state the results of our model – driving is cheaper, overall – and interpret it as a recommendation to drive to work today. A reality check (3d) for this recommendation might be the realization that recently there has been a construction project doubling your drive time. Or maybe your daughter had the sniffles last night and there is a good chance that your spouse might need to pick her up from school; the expense of a cab or ride share can change the equation for *today*. If you discover overlooked factors, you will have to re-estimate your costs and run your model again. Once you are sure there are no red flags, you conclude that the recommendation to drive is sound, and you (3e) write a summary of the analysis.

The chapters generally follow the template above, though not formally. Instead, the text flows more freely, breaking along the way to review the mathematical components of our argument. The topical examples from a wide scope of fields should be of interest in their own right but also demonstrate a great range of analytical methods and techniques in action – guiding the reader toward mastery of quantitative reasoning.

1

Is College Worth the Cost?

Appendix Skills: Arithmetic

OVERVIEW

We first react to the question, gauging our emotional response and recording the various ways it can be interpreted. After settling on an economic analysis, we identify the data we need, gather them, and compare college versus no college. In brief, we have the following steps:[1]

1 *Frame the question: here we decide to decide it on financial grounds, noting how this may be insufficient for many purposes.*
2 *Identify the costs and benefits of college, i.e. tuition and fees versus a good future income. Do the same for nongraduates.*
3 *Research the expenses and typical salaries of college graduates and nongraduates; discuss reliability of sources and limitations of the data we find.*
4 *Add up lifetime benefits of college versus no college, explicitly cataloging assumptions needed to use the information we could find. (One never has perfect data.)*
5 *We find a large lifetime benefit of college versus no college, leading us confidently to answer the question in the affirmative. We note that we have not decided the question of private college versus state school.*

Framing the Question

Gut Reaction

Is college worth it? Most families face this question explicitly or implicitly. What's your gut reaction?

> *Is college worth it?*
>
> *Think of your response and jot it down.*

Like many of the questions in this book, the topic is an emotional one. It may be difficult to give a clear-headed analysis after you and your family have been anguishing over the issue for years. Maybe you have been pressured to go to college

---

[1] The section headings in this chapter correspond to the template described in the introduction.

so that you can earn a decent living. Maybe your grandmother who takes care of you has been struggling lately, and you don't want to burden her with the tremendous expense. Maybe your friend went to a public university and his mother gave him a car with the money she saved versus private school. Maybe some of your friends aren't going to college, and just the thought of being away from them tugs at your heartstrings. Maybe your father expects you to attend his alma mater. Maybe your sister didn't go to college and she's doing fine – or maybe she struggled mightily.

Usually, some combination of these and other questions makes the college decision fraught with emotion. We can see that these emotions may fly in all different directions. That's okay; we're human – but we should be mindful not to let our opinions drive the analysis. For example, if you are taking this course in college, you have already made a choice and may be tempted to tip the scales in favor of your decision. Being aware of our emotions and potential biases will help insulate us from being swayed by them.

Clarifying the Question

Maybe you have no interest in money but simply love poetry. For you, it's not really a question of economics – the emotional "cost" of not studying poetry in college would be enormous.

To be concrete, let's decide that we are talking about the *financial* cost of college. This leaves open the question of *which* college. It becomes clear that there are too many colleges to answer the question specifically for each individual. We could do a separate analysis for each college and take some kind of average, or we could do a case study – a single, representative analysis for one particular college – then repeat it with different colleges/numbers as needed. Since this text emerged from a course designed for students at Northwestern University, we will select that college (in the year 2015) for our case study, clarifying the question to mean

Financially speaking, is it worth it to attend Northwestern University in 2015?

(Students from other schools will want to consider their own institutions instead of Northwestern.) This doesn't nail down the question completely – worth it for whom? Circumstances are different for different students, and not all have the same college costs. For example, some receive scholarships or grants.

We will encounter further questions along the way, but we now have a more clear conception of the question.

Remark 1.1 In fact, the driving question of our chapter lies at the heart of the field of *microeconomics*: an individual is presented with a choice and must weigh the different options so as to choose the "best" one. What is best will vary from person to person. Each choice also presents an *opportunity cost*, meaning the value of the best alternative: if you go to the movies on Friday night, it may be great fun, but you'll miss out on going anywhere else, e.g., to the symphony.   ▲

Envisioning an Answer

Now that we know the question, do we have an equally clear conception of the answer? Are we looking for a yes/no response, an up/down vote on the decision to matriculate? Or are we looking for an estimation of the net cost or benefit of attending Northwestern? Either way, we'll have to run the numbers, arrive at a figure, and then answer the question, so let's say that we'll go after the net cost/benefit of attending NU.

Now it is clear that if you live in the wilderness for four years versus paying hundreds of thousands of dollars for college, you will emerge in better financial shape after your four years of soul searching off the grid than a college grad who has plunked down many thousands in tuition – even if you do incur some expenses. However, the expectation is that a college education will land you a decent job that will pay off *down the road*. So we probably want to give an answer that tallies the costs versus benefits of attending Northwestern over the course of your lifetime.

Specifically, we will answer the question by giving a number representing the lifetime benefit, in dollars, of attending Northwestern. The number will be derived by adding up the benefits (such as the extra income conferred by a high-paying job) and subtracting expenses (e.g., tuition). Note that if benefits minus costs is a positive number, then college is worth the price.[2] But that's not all; we will compare with a similar analysis for the no-college option.

A cost/benefit analysis is a basic component of economic theory. In a sense, all decisions can be approached from this viewpoint: we compare the expected net gains of the different choices.

Building the Model

What Factors Will We Explore?

With this clearer view of the problem and the kind of response we wish to generate, let's start the brainstorming session. What factors will our analysis include?

> *What comes to mind?*
> *Think of your response and jot it down.*

> *What else?*
> *Think of your response and jot it down.*

Think some more. Reading my list is too passive. Creation is *active*.

> *Write at least one more thing.*

Here's my list, with the symbols + indicating revenues, − indicating costs, and ± for unknown.

---

[2] We could tabulate costs minus benefits, but then a negative value would mean an affirmative answer to the question, and that just seemed wrong.

- Tuition
- Room and board
- Books
+ Income, e.g., salary from job
- Lost income by not working throughout college ("opportunity cost")
+ Networking opportunities?
- Other hidden costs
± Overlooked/unknown factors

Your list and mine might differ. It's hard to guarantee that we've covered all the bases.

Data

Okay, now we need to do some research and find out the actual cost of tuition, room and board, etc. This will take some time: we will have a few rabbit holes to explore. This is our first chapter, so we need to give due consideration to each issue that arises.

In many cases, this is the hardest part of the process. Sometimes data are a proprietary secret (for instance, how much does Quaker pay each year to settle lawsuits from people who have chipped their teeth on hard raisins in granola cereals?) or simply too hard to find out (how many people does the government of North Korea employ to guard their Dear Leader?) or not available at all (how many hours each day did the average Aztec warrior spend on grooming?).

We will browse various sources for data, accepting only figures that are deemed *reliable, representative, and unbiased.* That is, we need our facts to be accurate, to pertain to our question, and to be free of prejudice.

This chapter would not be complete if I did not do this data gathering for you, but in the real world (or in the Exercises and Projects sections), you will be on your own. Students reading this chapter who attend other schools should try to find the analogous information for their own institutions (see Exercise 2).

In this case, a short Internet search reveals Northwestern's admissions website,[3] which details the costs for the 2015–16 academic year. The site reveals some of the other hidden costs that we didn't think of above. Still, our preparation has given us a good guide. Here is what the Northwestern site tallies:

- Tuition: $48,624
- Room and Board: $14,936
- Books and Supplies: $1,620
- Fees (health $200, student government $174, athletic $49, loan fee $35): $458
- Personal Expenses: $2,457
- Transportation: Varies
- Total Cost: $68,095

---

[3] The figures for 2015–16 are no longer available online. Figures as of this writing are available at http://admissions.northwestern.edu/tuition-aid/index.html.

Should we believe these values? Certainly, there is variation among students. For example, according to the website, commuting students have an estimated total cost about $12,000 less per year. So here we must clarify what kind of student we're talking about. Since most students who attend Northwestern are not commuters, we will not assume that the student is a commuter. Alternatively, we could have made a more complicated model, taking into account different contingencies: commuter or not, history major or engineer, pay tuition in full or take out loans.

Should we trust the numbers that Northwestern provides? As far as I know, colleges have not been known to misreport their own expense data. Some colleges – a handful – distort admissions data to *U.S. News and World Report*, which compiles a yearly ranking of schools. This does not appear to be commonplace, and the skewed figures involved were not as straightforward as standard fees. We will therefore consider these self-reported numbers as reliable.

With costs wrapped up, we turn to benefits. This will be trickier. Let's collect some data.

- Income. Note that attending college doesn't mean graduating, and of course not all graduates get a job, let alone earn equal pay. So we will have to take some sort of average value to build a simple model.[4]

  Here are some data one can find by following one's nose. The 2013 Salary Survey of the National Association of Colleges and Employers reports that the average starting salary of a four-year college graduate was $45,000. (*Forbes* carried the story.) So that's one number – but all colleges are different, and we are interested in Northwestern. Let's look some more.

  The company PayScale reports on its website that early career graduates of Northwestern University earn about $54,000 per year. These data are compiled from NU alumni who took the survey. Are they reliable? Maybe the high earners are more likely to want to take such a survey. With no way of knowing, we should worry that the respondents might not represent NU grads in general. This is what is known as *selection bias*: members of your sample may be predisposed toward a conclusion, such as if you go to Wisconsin to ask people whether or not they like brats. Furthermore, the whole PayScale survey involved 664 people, of whom 272 were early career (less than five years' experience). We should ask, *Is this a large enough sample for reliable data?*[5]

  A third source, the National Center for Education Statistics (the NCES is a government entity dedicated to collecting and analyzing data related to education), reported a median income of $48,500 for college graduates aged 25–34 in 2015.[6] This seems like good data from a reliable source, but the data are not special to Northwestern. In short, the (good) data we have are not quite the data we want,

---

[4] Different notions of averaging are discussed in Appendices 8.1 and 8.2.
[5] At this early stage of the course, we will not explore the issue of "small *n*" any further – see, e.g., Appendix 8.6.
[6] The site shows data for the current year: https://nces.ed.gov/fastfacts/display.asp?id=77.

but at least we have a few numbers from which to build a model – and the numbers are fairly close to one another.

We will wind up taking the NCES figure, since it is a government source (reliable, unbiased) and still in line with the other sources, which may be more specific.

- Lost income/opportunity cost. *Opportunity cost* refers to money not earned by choosing an alternate path. By going to college rather than working (if working were the alternative), a person fails to earn some amount of money, and this money can be interpreted as a further "cost" of attending college relative to another option. We will use the data collected here to evaluate the net benefit of not attending college, so that we may compare.

  So how much would you have earned had you not attended college? The naïve thing to do would be to look up the average salary of US workers and use that figure. But we must be more careful. We're not talking about a mid-career lawyer, we're talking about a recent high school graduate, so we must try to narrow our search to reflect the correct demographic.

  How do we search? No doubt any reader will just begin by Internet searches with the obvious keywords. This will pull up all sorts of sites making all sorts of claims, and way too many pop-up ads (a sign: commercial interests might mean a biased party). Beware: most of these sites did not collect the data they are reporting. However, even the only marginally responsible ones will report the source of the data. Click through to the source: you may find that many different media outlets/blogs/sites link to a common origin, and this is likely to be more reliable. ("All roads lead to Rome.") In this case, after a little browsing, I came upon the US Bureau of Labor Statistics (www.bls.gov), a neutral and reliable entity dedicated to gathering just the information I was seeking. At the site, I searched for "salary of high school graduates" and found a 2014 chart titled "Earnings and Unemployment Rates by Educational Attainment."[7] The data gave me more than I bargained for, but under "high school diploma" are a couple of key figures: high school graduates have median weekly earnings of $668. A quick conversion (see Appendix 5) yields the equivalent annual income:

$$\$668/\text{wk} = \$668/\text{wk} \times 50\text{wk}/\text{yr} = \$33,400/\text{yr},$$

where we have assumed 50 working weeks in a year. This chart also reported that high school graduate workers have an unemployment rate of 6.0%, while only 3.5% of workers with a bachelor's degree are unemployed. There was also a note:– "data are for persons age 25 and over" – as well as a link to another valued source, the US Census Bureau (www.census.gov). To get a better figure, at BLS, I searched for "recent high school graduates" and found a useful chart about how many of them had entered the workforce (but not their salaries). I found the following paragraph:

---

[7] www.bls.gov/emp/ep_chart_001.htm.

Recent high school graduates not enrolled in college in the fall of 2013 were more likely than enrolled graduates to be in the labor force (74.2 percent compared with 34.1 percent). The unemployment rate for high school graduates not enrolled in college was 30.9 percent, compared with 20.2 percent for graduates enrolled in college.

These unemployment figures mean that any assumption of full employment will introduce errors, in particular skewing the non–college graduates more.

According to the National Center for Education Statistics again, the median annual income of full-time workers age 25–34 with only a high school degree was $30,000 in 2013. This is quite close to the BLS number, which we found to be $33,400/yr. It is possibly lower due to the fact that the NCES only considered workers aged 25–34, but there is no need to speculate at this stage.

Happy I had enough for now, I bookmarked the site and moved to the next task.

- Income through networking. Do college graduates make money through connections they made while in college? Do these connections translate into more income compared to the connections that non–college graduates make in their lives? Such connections are often pejoratively called the "Old Boy Network," and rightfully so. But let's face it, we all have friends, and most of us listen to them, no matter if we're old, or boys, or stand to profit from their help. We needn't moralize here (though we should note any biases we carry), as we're more interested in the question of how we can measure the effects of networking.

  If I search for "measuring the benefits of the old boys network," an academic article from 2011 comes up first.[8] It's about the pay gap (about 30%) between male and female executives' salaries. According to the first page (I didn't pursue it too deeply), "We find that executive men's salaries are an increasing function of the number of such individuals they have encountered in the past while women's are not. Controlling for this discrepancy, there is no longer a significant gender gap among executives." Is this interesting? Yes! First of all, women get paid less than men, an important fact that needs to be explored and not dismissed. Second of all, there is indication that the difference could be attributed to networking differences between the sexes – so this, too, must be understood. Questions of access, opportunity and fairness arise and deserve careful investigation – and these would make an excellent project for this course!

  That said, we should return to the question at hand.[9] Is this new information useful for us now? Maybe. It does tell of an impact of networking on pay, but not between high school versus college graduates. I also searched for "benefits old boys network on income" and found a paper measuring the differential benefits

---

[8] The Old Boy Network: Gender differences in the impact of social networks on remuneration in top executive jobs, by Marie Lalanne and Paul Seabright, http://idei.fr/doc/wp/2011/gend_diff_top_executives.pdf.

[9] Or not. Maybe your academic explorations unearth a "drop-everything-now" discovery and you change the focus of your research. But we can't write a book that way!

between white male networks and those among females or minority-member networks.[10] These papers suggest there is an effect of networking, but don't lead us any closer to measuring that effect, and don't tell us whether it favors high school graduates or college graduates. Unfortunately, we simply don't have data relevant to our question and will have to punt on this point for now.

Remark 1.2 When I went to my college reunion, I was quite surprised (based on cars, clothes, size of the class gift, etc.) at how much money everyone must have had. I've heard that wealth begets wealth, and it didn't seem that such opulence could be attributed entirely to the benefits of college. It may have come from other connections: family ties, friends in high places, etc. But despite our hunches, we cannot proceed without good data.   ▲

- Other? It "feels" odd that the only financial benefits we have identified for not going to college are just saving money and getting an early start at work. What about all those Internet billionaires who dropped out? Is their only advantage that they started up their companies that much earlier? (That would fall under opportunity cost.) What about the environment of geniuses that they found in Silicon Valley? (That would fall under networking benefits, though we decided above that we did not know how to quantify this.) What about the freedom to roam and explore in thought and deed? To be intellectually free and not bound by traditional college compartmentalization? Are nongraduates *better* equipped? Maybe all of this is true – and if it is, I hope non–college students are also reading this book! – but we must hope that these benefits, if they exist, are measured in the income of non–college graduates as tracked by the Bureau of Labor Statistics. If they are somehow "dormant" or "intangible" benefits, we will not be considering them here. That's the decision we made when we decided how to interpret our central question.

  Still, there is always the chance that we may have missed something. We'll have to listen to the critics after publishing our results.

There are other data that we may need to collect down the road, but already we know that we plan on adding up the *lifetime* cost or benefit of college, so we'll have to figure out what a lifetime of work means. We won't assume that we work up until our life's end, but it will be useful to know what the life expectancy is to estimate our working life-span. I looked on Wikipedia and found a few lists of countries and life expectancies but wanted a better source, so I clicked through and found the CIA World Factbook,[11] which stated that the life expectancy in the US is about 79.5 years, so we might want to plan for retirement based on that age.

Of course, retirement age varies from person to person, and there are laws that factor into the decision. Social Security acts as a federal pension program, and if

---

[10] Steve McDonald, *What's in the "old boys" network? Accessing social capital in gendered and racialized networks,* Social Networks 33 (2011) 317–30, www.academia.edu/13356204/ Whats_in_the_old_boys_network_Accessing_social_capital_in_gendered_and_racialized_networks.

[11] www.cia.gov/library/publications/the-world-factbook/rankorder/2102rank.html.

you were born after 1960, you can collect retirement income without penalty from age 67.[12] Researchers have found that the average retirement age is 62 for women and 64 for men.[13] Will it still be 62 when today's female college graduates retire? Who knows? The average retirement age is rising, and there is pressure to raise the age at which one can collect Social Security so the coffers don't go empty. Here we will *assume* that the average retirement age when today's college students become more senior will be 67 years.

These rules and programs, and people's behavior, may change by the time someone entering college now is ready for retirement, but the best we can do is plan based on what we see today – so let us assume that retirement begins at age 67.

We will also have to say something about the age of the student. According to Northwestern's website, 86% of students graduate in four years and 92% graduate in five years. Note this represents a small but appreciable chance that a student won't graduate. Dropping out will lower your expenses, but you won't have the benefit of a graduate's higher salary. There were no data about age of students, so we will simply assume the student begins after high school at age 18. Of course, if you are building your own model, you may tailor all the numbers, using your own age and the expected number of years you plan to be in school. We will assume that students going to college will graduate, recognizing that another assumption introduces another source of error.

Now we can decide what to call a lifetime of working. If most college graduates graduate at age 22 and work until they retire at age 67, then a lifetime of postcollege work is 45 years – or 49 years for someone who forgoes four years of college.

Assumptions

We are almost in a position to construct a model of our answers. A naïve answer would just add up benefits and subtract costs, but we have realized it's not so simple. We must make some assumptions about the person attending college (a noncommuting student, we have decided) and whether we want to use the data about "average" salary or narrow our focus to a more specific demographic. So if we are a female student, should we focus on earnings among female high school graduates, for example?

We don't yet know how much of the information we have gathered will be used in our model, but for clarity, we record a concrete list of assumptions.

- We will use the Northwestern numbers for college expenses.
- For the salary of recent Northwestern graduates, we will take the data from NCES and assume a $48,500 starting salary. This is clearly a rough guess, since it does not represent any added benefit of going to Northwestern rather than a different school. However, it is within roughly 5% of the other two numbers we found (the

---

[12] www.ssa.gov/planners/retire/agereduction.html.
[13] Alicia Munnell, *What is the average retirement age?*, http://crr.bc.edu/wp-content/uploads/2011/08/IB_11-11-508.pdf and *The average retirement age – an update*, http://crr.bc.edu/wp-content/uploads/2015/03/IB_15-4_508_rev.pdf.

higher of which we suspected to be an overestimate), so we can expect that this estimate introduces a margin of error into our result of $\pm 5\%$.

- We will use the NCES data for income of recent high school graduates and the BLS data on unemployment rates.
- We will assume 45 years of work for a college graduate and 49 years for one who does not attend college.
- We will use the Northwestern numbers concerning degree times and degree completion rates.

Constructing the Model

Okay, now how are we going to add things up?

We have the cost of college, but we know that if we can't pay the cost up front, then we'll have to take out a loan – and interest on that loan will be a hidden cost. (We will explore loans and interest in depth in Chapter 3.)

Adding up the salary is much harder, as it changes over a person's lifetime, and one simple figure probably won't do. A rough guess may be good for a blog post, while a precise estimate could be the content of a PhD thesis. We're not aiming for a PhD (yet!), so we'll be content with something in between, starting with a basic calculation.

A First Approximation

Let us run the numbers on a simple model. We will discuss improvements shortly afterward.

For college graduates, we can simply multiply our assumed income of $48,500 per year times the number of years of working after college, assumed 45. Simple multiplication then gives a total earnings of $2.18 million. Four years of college at Northwestern, we decided, costs $4 \times \$68,095$, or about $272,000.[14] Subtracting the costs from the revenue gives a lifetime college benefit of $\$2,180,000 - \$272,000 \approx \$1.9$ million.

For earnings by high school graduates, recall the NCES datum of $30,000/yr. Someone who enters the workforce at age 18 and works for 49 years will then earn $1.47 million.

So college graduates come out ahead by about $1.9 million, while high school graduates come out ahead by $1.5 million. This puts the value of a college education at about $0.4\,M = \$400,000$.

Remark 1.3 Ballpark it! Even with little or no research, we probably could have crudely estimated that college costs, say, $250,000, and that graduates earn $50,000

---

[14] In Appendix 5.3 we discuss how to decide how many digits to keep. Giving a number with the precision of $68,095 when we have already acknowledged that this can be off by quite a bit is silly. Or, even if I know some number precisely, when I add it to another that I only know vaguely, I can't know the sum precisely. So we keep only a set number of significant digits. Put simply, we round.

a year for 45 years, giving a lifetime net of $45 \times \$50,000 - \$250,000 = \$2\,\mathrm{M}$, and that nongraduates make something like $50 \times \$30,000 = \$1.5\,\mathrm{M}$. ▲

Improvements

Our first guess of the economic benefits of a college education, while not very reliable, is noticeably large. This hints that the general conclusion that college is worth it should be pretty robust, so that after a finer analysis, we may find different numbers but the same recommendation: go to college.

If we were to make a finer analysis, we'd need more data – see Remark 1.4. Here are some items to consider:

- Interest on the student loan. We'll want to know what the size of the loan is and what the interest is. Data for Northwestern or comparable schools will be preferred. How would we add this feature to our model? Well, using what we might know about accrual of interest (see Chapter 3), we could calculate the total cost of a loan for college rather than just the sum of tuition and fees. Such an analysis might assume a fixed-rate loan, but more sophisticated loan schemes could be investigated as well.
- Change of salary over time. Typically, workers get periodic raises equal to some percentage of their salary. (This would mean that your salary grows exponentially over time – see Appendix 6.4.) But we have modeled salary as a constant function, a fixed number – $48,500 – and even if that number is true today, it represents an average of people at various ages. Who's to say that today's figure for a 65-year-old will represent the reality 47 years from now, when someone who is 18 today turns 65?

We have gathered some data on student loans. According to the Institute for College Access and Success (TICAS), which conducts nonpartisan research, student loan debt at graduation in 2015 averaged $30,100 in the form of a ten-year federal loan with a fixed interest rate of 3.76%.[15] (In Chapter 3 we will discuss loans and interest and what all this jargon really means.) Now not all students have debt, but TICAS found that seven in ten graduates of nonprofit colleges did, so we will make the assumption that we will need to pay off a $30,100 loan at 3.76% interest. As we will learn in Chapter 3, such a payment plan would mean monthly payments of about $300, for a total cost over the ten years of $36,000, including interest of about $6,000. We can already see that this relatively small extra amount won't affect our conclusion.

Let us turn to the fact that salaries change over time. The BLS[16] provides data (from 2015) on average weekly earnings of full-time workers by age, adjusted here assuming a 50-week work year to represent annual income (Table 1.1).

---

[15] *Student debt and the class of 2015,* Institute for College Access and Success, https://ticas.org/sites/default/files/pub_files/classof2015.pdf. We note that not all debts were from federal loans, and interest rates varied, but for simplicity we assume here a federal loan at 3.76%.
[16] www.bls.gov/news.release/wkyeng.t03.htm.

Table 1.1 US Median income by age

Age	Median Income
16–24	$24,000
25–34	$36,800
35–44	$44,650
45–54	$46,500
55–64	$45,150
65+	$41,900

We will use the data in Table 1.1 to figure out expected incomes for college graduates and nonattenders. Since we only have overall data, we will have to make the assumption that the median income follows the same pattern – that graduates and nongraduates alike get *percentage* raises equal to the raises on this chart. As we will find out in the section "Calculation," this will imply that the whole table scales up or down based on the different starting salaries.

Oops! Missed Something!

In the first draft of this chapter, I made an error. I have left it in for instructive purposes, but if I were creating a report for work, I'd have to do a full rewrite. Did you spot it? The problem is that we have rolled room and board into the expenses of college but did not consider that a noncollegian would also need to pay for food and housing. We can immediately see that this would only widen the earnings gap between college and no college, so the mistake would not affect the conclusion. We can look more closely and estimate room and board for four years of noncollege life. Even at $1,000 per month for 48 months, we're talking about less than $50,000, and based on our first approximation, this should not impact our conclusion significantly, but a potential 10% (or so) correction is important. Or, we could assume that our noncollegian lives at home to save money (of course, the food expense is incurred by *someone*). We will take this later route, because it allows the written record to reflect the complicated chronology of our reasoning. One way or another, we of course needed to address our omission and fix the error. This brings up a general point that is worth underscoring.

Remark 1.4 *Loops in our flow chart.* It would be convenient to label the steps – build model; gather data; calculate – and to write this chapter as if the process were so tidy, but in reality the chronology is often not so simple, as we have seen here. Sometimes in gathering data, you encounter evidence that drives the model. Sometimes at the modeling step, you realize you have insufficient data for a model with the sophistication you desire. Sometimes, after investing lots of time and energy, you are dissatisfied and find yourself demanding an even better model. (Beware: this is how academicians are made!) Sometimes, the brass tacks of a calculation reveal gaps that need filling. We all want a polished result, but this will require smoothing some bumps. ▲

Sources of Error

There are many sources of error here. Try to think of some.

List them here.

Here's a list I came up with.

- Salary levels vary widely from person to person and over a single person's lifetime in a way we cannot precisely model (with the data at hand).
- Salary data were given in median, rather than average (mean), values.
- The value of money changes over time.
- Costs of college will vary depending on the terms of a student's loan.
- We have not accounted for the cost of job training for non–college degree recipients.
- Higher earners can save and invest, leading to further increases in earnings (or losses).
- All of our numbers for nongraduate earnings are averages across all such workers. However, to make our comparison fair and consistent with the fact that we are considering the question "Should I go to Northwestern?" we should only compare Northwestern graduates to nongraduates *who could have gone to Northwestern but chose not to.* Such people might have different earnings than high school graduates who were not admitted to college.
- Unemployment rates are higher among nongraduates (6%) compared to workers with a bachelor's degree (3.5%).[17]
- About 8% of Northwestern students do not graduate. These students would still pay for the cost of some college, but potentially not receive the economic benefits. To account for this, we could use probability theory to calculate the expected income after enrolling in college, including the possibility of not graduating.[18] We could further fine-tune if we had data for salary for those with *some* college experience.
- We have assumed that the noncollegian lives at home for four years, spending no money on room and board – but this person is then dependent on someone, who still incurs the costs that we ignore.

As we shall see below, the conclusion that we draw will be robust enough to argue that these sources of error will not affect our answer. (We've seen hints of this already.) Had the result been marginal, the errors cited above could play a pivotal role. For example, higher earners may be able to add to their income more easily ("the rich get richer"), but such a factor would only *increase* the strength of our conclusion. Or, a wage earner could potentially invest while the college student

---

[17] www.bls.gov/emp/ep_chart_001.htm.

[18] See Appendix 7.1 for a discussion of expected values. In the present case, the expected starting salary, for example, would be calculated as $0.92 \cdot \$48,500 + 0.08 \cdot \$30,000 = \$47,000$. The first term represents the chance of graduating and getting the college-graduate salary, while the second term represents the chance of not graduating and receiving the lower salary.

is attending school – but to make up such a large deficit would take an extremely savvy (or lucky) investor. On a similar note, the value of money changes over time, so it is wrong to treat future dollars on an equal footing with present dollars. As another example, the average salary is likely to be higher than the median, since high earners make very many times the median, while even people who earn nothing at all make only one median less than the median. Using average salaries, then, would likely only have exaggerated the differences between college graduates and nonattenders.

Calculation

Let's get cracking on the improved model, including a graduated salary scale and the cost of interest on student debt.

With interest payments on the student loan coming in at about $6,000, we must add that to the cost ($272,000) of college. This difference will turn out to be negligible compared to the degree of precision we will employ, so in fact this consideration will ultimately have no effect on our analysis. But for now, we add the two costs, keep two significant digits, and record the total cost as the sum $280,000.[19]

How will we incorporate the graduated income levels of Table 1.1? Unfortunately, we don't have these data for the cohort of graduates versus the nongraduates. However, we can make a simple adjustment. If, for example, Table 1.1 had indicated that salaries tripled after every decade, we could apply the same rule for graduates and nongraduates alike, and the different starting incomes would mean different graduated lifetime incomes. If, say, the college graduate made *twice* what the median worker did at a similar age for which we have data, and worked the same number of years, then by doubling that higher number each decade, we'd see that after a lifetime, the college graduate would earn *twice* (the same factor) what the median worker did.

To make the numbers simpler, suppose that the median worker earned $1, 3, 9, 27$ in four subsequent decades (ignoring units, for illustrative purposes), tripling with each raise. We'd then get a total of $10 + 30 + 90 + 270 = 400$ for the four decades, as the numbers each get multiplied by 10 since they represent salaries for ten years. Now if a college graduate started with twice the income, 2, and tripled each decade, that person would have incomes of $2, 6, 18, 54$ for a total of $20+60+180+540 = 800$ for the four decades, precisely twice![20]

Likewise, if nongraduates earn *three-fifths* of what the median workers do at the start of their careers, then they would earn *three-fifths* (the same factor) of what the median workers earn over a lifetime. So we can use Table 1.1 to calculate median

---

[19] Significant digits are discussed in Appendix 5.3.

[20] This is an exercise in the distributive rule – see Appendix 2.1. If a salary of s gets raised by a factor r (so $r = 3$ for tripling), then the total for two years is $s + rs$. If a salary of $2s$ gets raised by a factor r, then the total for two years is $2s + r(2s) = 2(s + rs)$, using distributivity – giving *twice* $s + rs$. Now note that the 2 may be replaced with any factor a, giving $as + ras = a(s + rs)$.

income and then multiply the answer by the ratio of our cohort's earnings to the median worker at a similar age for which we have data.

So we will first calculate what the median earnings are for the typical worker working from 22 to 67 (for the college graduates, who we have assumed will start later), then do the same for the ages from 18 to 67 (to compare with the nongraduates, who we have assumed start work earlier). *Then*, we will scale these numbers up or down based on the ratio of earnings between our cohort and the median worker. The comparison will give us the lifetime differential in earned incomes for our two cohorts.

First, we consider a median worker who starts work at 22 years old. Such a person will spend two years in the 15–24 age bracket for a total earnings of $48,000, ten years in the 25–34 bracket for a total of $368,000, and so on. Continuing in this manner until age 67, we arrive at a total lifetime earnings of $1.86 million. If we assume that work starts at age 18 after high school instead, this number would increase by $4 \cdot \$24,000 = \$96,000$, to about $1.96 million.

Next we compare the numbers we know about starting income – $48,500 for college graduates, $30,000 for nonattenders, both in the 25–34 age range – against the median 25–34 age range value of $36,800, which includes both populations. These numbers determine the ratios – $\frac{\$48,500}{\$36,800}$ and $\frac{\$30,000}{\$36,800}$, respectively – by which we must multiply the lifetime earnings of the median worker, which does not account for knowledge about schooling, to get data for graduates and nongraduates.

So the Northwestern graduate is estimated to earn

$$\frac{\$48,500}{\$36,800} \cdot \$1.86\,\text{M} \approx \$2.45\,\text{M}$$

but pay about $0.28 million for school, giving a difference of about $2.2 million. The college nonattender is estimated to earn

$$\frac{\$30,000}{\$36,800} \cdot \$1.96\,\text{M} \approx \$1.60\,\text{M},$$

a figure we record as $1.6 million. The revised benefit of college is then now $0.6 million, or $600,000.

So the revision, since the wage gap only broadened over time, significantly strengthened our conclusions about the benefits of college.

Alternate Model

Our analysis isn't the only approach we might have developed. For example, we could have looked at the problem from a question of cash flow. Given that a college graduate would have a student loan to pay off and a monthly payment on that loan, would the difference between the person's salary and the salary of a high school graduate be more than the cost to service the loan?

We can answer this question right away. We saw that the yearly payments on loan debt, at $12 \times \$300 = \$3,600$, were not a large fraction of the excess salary ($18,500) that a college graduate earns. Had loan payments been higher than the salary differential, then college graduates might be worse off than their nongraduating peers, at least initially after college. So even if the answer to the question of the total benefit of attending college is "eventually, yes," if the loan were not serviceable upon graduation, then we would have a real problem.

In the present case, due to the relatively low cost of servicing the debt compared to income, even just out of school, we find that a cash flow model would not affect the take-away from our analysis. (Many students have different stories to tell when it comes to debt! There is no one-size-fits-all argument.) However, in other scenarios, such as companies investing large start-up costs to arrive at an eventual profit, cash flow considerations might lead to dramatically different conclusions.

Analyzing the Results

Check Work!

Check work. If you are reading this, it has probably passed through many sets of eyes before appearing on the page/screen in front of you. (The author had help.) But in real life, we have to stand on our own two feet. Tips for checking your work:

1 Re-do it another way. For example, when I grade exams and add up the scores on five test problems, I'll tally up the scores from questions 1 through 5, then tally questions 5 through 1 to compare. This simple trick can help catch errors. You can use a variant for your own work.
2 Get a ballpark estimate with a back-of-the-envelope calculation, as a reality check – see Remark 1.3 and Appendix 9.
3 Run the simulation/model with different numbers to make sure the result seems sensible. I could make the cost of college $3 billion, and this should certainly expose an error if college is still seen to be worth it. If something like this happened, maybe we had transcribed some "maximum" salary among college graduates rather than an average salary, or maybe we just omitted the college cost from our spreadsheet. Either way, running a few numbers to make sure we get sensible results is wise.

Precision?

How certain are we of our outcome? In terms of numbers, there were too many sources of error to be confident in the precise number. However, the fact that the number was large and positive indicates a strong net benefit to attend college versus no college at all – and gives a robust, affirmative response to the question of whether attending is worth it. This is a conclusion we can make confidently.

What Does Our Analysis Tell Us?

The great physicist Richard Feynman, who developed a formulation of quantum mechanics, said,

> When you are solving a problem, don't worry. Now, after you have solved the problem, then that's the time to worry.

Now we have to interpret our result and come up with a sensible recommendation. Here the decision is clear: the large net benefit means we can answer the question. Yes, Northwestern is worth it.

We saw significant potential errors, but the results came back resoundingly clear that we can conclude with a high degree of certainty that for the typical student (as defined above), attending Northwestern University is worth the cost. Our expected net benefit of $600,000 more in lifetime earnings versus not attending college is a number fraught with too many potential errors to bandy about, yet we can be confident that in most cases, the number will be large, in the hundreds of thousands of dollars.

A very relevant question we did not answer was whether it makes more sense to go to Northwestern versus, say, a lower-priced state university. We only asked about Northwestern or nothing. Answering the more pointed question would require looking at variations in income levels for college graduates across different schools. Surely there are some surprises yet to be discovered.

Addressing this subtler issue or trying to give a precise numerical benefit with more certainty would require a much finer analysis and more specific data. (Our reluctance to assert a specific number for the benefit of college doesn't mean that our reasoning is not quantitative!) This can be done, but we're not after a PhD just yet. How could we be? We've only just decided on attending college!

Reality Check

The conclusion is a bit unexpected, given the high price tag of college, about a quarter million dollars! But given a long enough expected range of salary-earning capabilities (i.e., assuming that you live past a retirement age of 67), the money is earned back through higher income. Note that the conclusion may not apply to you if, say, you are thinking about starting school at age 60. Nor might it if you *already* have a high-paying job but are thinking of quitting to begin attending Northwestern to study philosophy. Anyway, in these and other cases, financial impact may not be your primary consideration.

Summary

Let's highlight what we did. We asked a question about the value of college and then decided to answer it on purely economic grounds. This was myopic. College can be many things to many people. To some, it is a heavy financial burden and major cause of stress, regardless of the eventual economic outlook. To others, it is a life-changing experience, a carefree time surrounded by friends and marked by personal growth. To some, it is where professional skills are developed. The value of the experience is a richly textured concept, much deeper than just the economic impact: students develop independence, learn to view the world in different ways, meet people from other backgrounds, and challenge themselves in novel ways. The true value of college is an extremely personal question, but not the one we considered. We studied only financial considerations.

To analyze the costs and benefits, we scoured the Web for data, focused on sources we deemed reliable, and built a model to tabulate the lifetime cost–benefit differential for graduates and nongraduates alike. We found that the increased income that graduates earn relative to nongraduates more than made up for the cost of university. We then considered sources of error in our calculations and determined that our conclusion was robust – different approaches and assumptions lead to the same conclusion, though with different numbers – even though we were unable confidently to attach a specific number to the economic value of college.

I hope this lesson was worth it. If not, at least I'm confident that the rest of college will be!

Exercises

1 Calculate from the data in Table 1.1 the expected lifetime benefit of dropping out of college to join a tech company at age 20 with a starting salary of $65,000. Assume that this person paid for half the price of a college education and will retire at age 67.

2 Run the numbers of the model for this chapter for the year and institution that *you* are attending right now. Do your own research and cite sources. (This exercise involves research.)

3 You are opening a shoelace factory and will need to take out loans at an annual interest rate of 5% to pay for construction and equipment. You will need $4 million to buy the land and build the factory, plus $80,000 for every 1 million shoelaces per year you produce. You know you can make a net profit of $0.10 per shoelace. You are confident that your profit margins will grow with time, so all you need to do to is ensure that you can pay the interest on your loan after the first year. How many shoelaces do you need to make in Year 1 to ensure a positive cash flow?

4 What is the cost of one college lecture? Explain your computation and comment on how worthwhile the number you obtain actually is. (This exercise involves research.)

5 Using our simpler first model that did not take into account changes in income over time, find the break-even point, i.e. the age at which the net benefit of attending college becomes positive. (Assume, as we did, a constant salary of $48,500 for college graduates and $30,000 for nongraduates over their entire career.) If we did take changing income into account, would the break-even point come earlier or later?

6 What is a "PILOT" (payment in lieu of taxes)? Identify some issues around tax policy for universities. (This exercise involves research.)

7 Think of a topic for a project related to this question.

Projects

A Construct a model to analyze the question based on cash flow, addressing, in particular, servicing student loan debt in the years immediately following graduation.

B Is it worth it to go to Northwestern versus a public state school? Pick one, develop a model, and run the numbers.

C News articles will tell you that it takes about 2,000 gallons of water to produce a pound of almonds, while 500 gallons is needed to produce a pound of chicken meat. How are these numbers figured? And is a pound of chicken versus a pound of almonds the right comparison to make? After addressing these questions, think of how you would perform the analysis, then run your own model, adding up all water costs of the processes involved in production.

2

How Many People Died in the Civil War?

Appendix Skills: Arithmetic

OVERVIEW

After discussing the value of estimating war deaths, we review the historian David Hacker's analysis by reviewing census data patterns and finding a deficit of males after the Civil War, compared to what one would expect from demographic trends. The analyses require careful assumptions, and we spend time reviewing what is needed to do so:

1 *We argue that with changing populations, a meaningful sense of a war's impact comes from finding, rather than an absolute number, the fraction of the population that dies.*
2 *We discuss the difficulty in determining which deaths are attributable to war, and difficulties in estimating the number of deaths from a war fought long ago.*
3 *We review the method of estimating deaths through census counts. This amounts to a more sophisticated version of the following: if some population in consecutive years follows the pattern 1,2,3,4,3, then you can reasonably deduce that something happened to hinder population growth in that fifth year.*
4 *Going from "something happened" to blaming the deficit on war requires arguments that nothing else was responsible; we therefore spend time reviewing assumptions.*
5 *After all this, we run the numbers and explain Hacker's calculation.*

Reaction

How do you even begin to think about such a question? And why? Is it even important? Isn't *every* war death a tragedy? The war was long ago; what can the number tell us today?

Sometimes the questions we confront arise organically, such as when you buy a can of mustard greens for just 99 cents, note that it was imported from India, and wonder *how on earth could that possibly be profitable?* Answering this would tell you something about global trade and labor. But sometimes a question just comes out of the blue, and we should then consider, even if we could find an answer, what would it mean?

A 2012 news story on the question[1] refers to an updated estimate of the answer appearing in the article "A Census-Based Count of the Civil War Dead," by J. David Hacker, published in the journal *Civil War History*. The story begins,

> The U.S. Civil War was incontrovertibly the bloodiest, most devastating conflict in American history, and it remains unknown – and unknowable – exactly how many men died in Union and Confederate uniform.
>
> Now, it appears a long-held estimate of the war's death toll could have undercounted the dead by as many as 130,000. That is 21% of the earlier estimate – and more than twice the total US dead in Vietnam.

Here we immediately encounter the importance of quantitative reasoning even in a nonnumeric field like history. Let's read again what is written. A new accounting of deaths has concluded that just the *undercounting* of soldier casualties,[2] namely, 130,000, was more than twice the total number of US casualties in Vietnam. The absolute number of casualties itself, 750,000, is much greater than that. These numbers are meaningful in trying to compare the relative significance and impact of two events that were not both experienced by any living person. Vietnam is somewhat more recent, but its US soldier death toll – measured quite carefully by the US Army – was about 58,000 (many others died from other countries). By comparison, the two Iraq wars and the War in Afghanistan have resulted in 6,757 US Army casualties as of this writing.[3]

Meaning in Numbers

Before turning to the Civil War alone, let us make some quantitative remarks about the relative impact of a few different wars fought by Americans. Very roughly from the numbers above, the Vietnam War had about nine times the number of casualties of the combined Iraq and Afghanistan wars – or more like six if we include the roughly 3,000 who perished in the 9/11 attacks. The wars in Iraq and Afghanistan are contemporary, so most of us, whatever our age, have some familiarity with their impact. So if we were not alive during the Vietnam War, we might try to extrapolate what nine or six times this impact might be like. How many more stories close to home would there be? How many more tales of the loss of a cousin, a friend, or a friend of a friend? It's hard to imagine the scale of the loss for those of us who have not lived through it. But now consider: the Civil War had more than ten times the number of casualties as Vietnam. Add to this the fact that it was fought on American soil and we might only try to begin to feel the magnitude of this event.

All of the preceding is meant to illustrate that the *number* we attach to the death toll will be of great importance to this and all future generations of Americans.

---

[1] www.bbc.com/news/magazine-17604991.

[2] The word *casualty* could refer to death or serious injury; we take it to mean the former.

[3] http://archive.defense.gov/news/casualty.pdf and www.archives.gov/research/military/vietnam-war/casualty-statistics.html#intro.

The conclusion that the number is important is certainly valid, but what about our analysis behind the number, our attempts to understand the impact? Does ten times the number of deaths mean ten times the impact?

Write your opinion here.

To begin with, we briefly consider a proxy question involving animals. This is only for illustrative purposes. It is certainly not meant to compare animals to humans. Instead, it will allow us to isolate an issue before returning to the matter at hand.

Suppose a rancher slaughters three angus cows at the abattoir.[4] Let's compare this act with killing three northern white rhinos. Both species are large herbivorous mammals. Are these events comparable in their impact? Perhaps it is relevant that there are an estimated 1.45 billion cattle worldwide,[5] while there are currently (as of this writing) only FIVE white rhinos. In fact, when one died recently, it was serious news![6]

So, without diminishing the loss of life, we may conclude that the *ratio* of the number of deaths to the population at large is a better measure of impact. In terms of ratios, then, three deaths among 1,000 people would be counted the same as six deaths among 2,000. This makes some sense, as we might imagine the 2,000 people inhabiting two separate small villages of 1,000 people, each of which lost 3 of its members. To use this more meaningful measurement – percentage of the population killed in armed conflict – let us compare the census data from the time of the Civil War to the time of the Vietnam War.

Remark 2.1 Always be on guard for whether an absolute or relative figure is more meaningful. If someone complains that the incidence of crimes is rising, probe to find out if there is an increase in the crime *rate*, namely, the number of crimes in a given period divided by the population. If the population is growing rapidly, the incidence of crime could increase while the rate is decreasing; newcomers may not be to blame.   ▲

The US Constitution requires that the census be taken every 10 years: in 1860 the population was about 31.4 million, while in 1870 it was about 38.6 million. Compare with the population in the Vietnam era: about 203 million in 1970. So how does the Vietnam ratio of 58,000 deaths out of 203,000,000 Americans (or 0.03%) compare

---

[4] This scenario may seem offensive or cruel to vegetarians or Hindus or other people. We could choose a different example, but continuing with the present one can be instructive. After all, this chapter concerns human death during war, a most difficult topic by any measure. It is a challenge to reason dispassionately about topics that are upsetting, but it is a necessary skill in life and in many professions. We will try to do so sensitively.

[5] www.fao.org/faostat/en/#data/QA

[6] See www.washingtonpost.com/news/speaking-of-science/wp/2014/12/15/a-northern-white-rhino-has-died-there-are-now-five-left-in-the-entire-world/. Since this text was first written, two more have died – see www.bbc.com/news/world-us-canada-34897767.

with the Civil War ratio of 750,000 deaths out of the population in 1861 of about 31.4 million (or 2.4%)? The Civil War ratio is actually 80 times as large!

Exercise: Give a whole number N such that 2.4% is about one in N.

Think of your response and jot it down.

The exercise asks for N so that $0.024 = 1/N$, or $N = 1/0.024 = 41.666\ldots$, so the nearest whole number is 42. So our number of 2.4% means about 1 in 42 dead. At 1 in 42, it is hard to imagine a single family that wouldn't be immediately, profoundly impacted by death from the war. Thinking more deeply, the impact of the war is much greater than the death of soldiers. Many able-bodied workers were summoned from their jobs, meaning less farming and less productivity, so less food and income for those left behind. What about the impact on life in this sense? These are all valid questions that only begin to touch on the impact of war, but we will remain focused on our single driving question.

Still – how can we count a number of people who died so long ago? How can our numbers possibly be better than numbers constructed from that time?

How do you think the numbers were originally obtained?

Think of your response and jot it down.

What sources of inaccuracies might there have been?

Think of your response and jot it down.

According to Hacker's article, in post–Civil War America, an effort was made to account for the dead, but many records were destroyed. An indirect number was formed after gathering information from battlefield reports, regimental rolls, and widows and orphans. The number became one of those "time-honored" figures that keeps getting repeated, but the honor of time may not be reflective of any great wisdom: simply, no better figure had been constructed.

Hacker's count was also indirect but reflected the results of a massive, systematic government effort already mentioned in these notes: the US Census. Now Census numbers are as reliable an estimate of the population as one can find, but what does this have to do with those killed fighting? And what does "killed fighting" even mean? Does it include those who died later of their wounds or infections, or nonmilitary personnel killed in raids, or those who died of diseases contracted during service? We must *define our terms*. Hacker's count includes all of these, and he terms this "excess male deaths," measuring those occurring between 1860 and 1870 (the war was from 1861 to 1865).

Opinion: Is this is a good measure of the toll of the Civil War?

Think of your response and jot it down.

This measure has an obvious omission: all females! It is absurd to neglect the contributions and sacrifices that females made in combat and in other ways during wartime. As we shall see in the next section, Hacker's choice is not from callousness but because the raw number of females involved is so low compared to males. Notwithstanding the role of females, this measure of the toll *on males* is reasonable: a later death due to injury or a nonmilitary death in a raid is still a death *attributable* to war. (In Chapter 4 we will learn of attributable risk – see in particular Exercise 9 of that chapter. The deaths counted here would be reflected in the attributable risk of mortality from the war.)

Two-Census Estimates

What's funny about this pattern: 1,2,3,4,5,3,4,5,6,7? It increases by one at each step, except for one place. The pattern had a disruption. Noticing a disruption of patterns is the idea behind a two-census estimate: in the example of our chapter, an increase in the rate at which white males were dying will reflect the effects of the Civil War.

Consider the following simple scenario. Suppose you work for five weeks in a lab that breeds fruit flies. In the fifth week, there was a hazardous waste incident in the lab. Nothing else out of the ordinary happened, and you made sure to keep the flies fed and cared for according to protocol. You had measured the following numbers of flies:[7]

Week 1	Week 2	Week 3	Week 4	Week 5
10	13	17	22	24

From Week 1 to Week 2, the number of flies increased by 3, a percentage of $3/10 = 0.3 = 30\%$. The next week, it went up by a percentage $4/13 = 31\%$, then 29%. Based on this pattern of about 30% increase per week, we would expect another 7 or so flies in Week 5. But there were only two more flies added in Week 5 instead of seven. Given that the hazardous waste incident was the only thing out of the ordinary that occurred that week, we can reasonably attribute the deficit of 5 to the hazardous waste.

In the case of the Civil War, we will look at survival rates in the decades before, during, and after the war. Diminished rates mean excess mortality: more deaths. Now let us look more closely. First, know that the Census measures "cohorts," such as the cohort of men aged 20–24 in 1860. If we compare that number to the number of men aged 30–34 in 1870 (they would have aged 10 years in the interim), then the difference represents the number in this cohort who died in the 1860s.

Right?

Think of your response and jot it down.

Well, not exactly. There could have been immigration or emigration – people moving into or out of the population – which affected these numbers. (Births

---

[7] These numbers seem a bit low for fly populations – maybe ten should stand for ten *dozen* flies – but the lessons are clear.

wouldn't affect this cohort, since no man aged 32 could have been born in the previous decade!) However, if you can discount immigration, then you could make this calculation for each cohort involved in the war, add them up, and have your answer. Well, assuming that the two censuses themselves were accurate. That's the ideal-world scenario, but we don't live in an ideal world, so our estimate will only ever be a guess.

> *Does that make it useless?*
>
> *Think of your response and jot it down.*

Assumption and the Quantification of Error

A flat-out guess would be useless, but if we can somehow ensure that the errors are limited, then the accuracy of the guess can be *quantified*. Quantification of error is an important part of any kind of analytic endeavor: the real world is always messy, and estimations are always imperfect, so a reliable measure of confidence is needed. Only pure mathematics can be perfectly correct: $3+7 = 10$, period. Another source of error in mathematical modeling is assumption. Even if I say "there are seven people in the room; how many total fingers?" the initial guess of 70 relies on an assumption that each person has ten fingers. This assumption is *not* true in general, so our estimate may be off. Some people have no fingers and some have as many as seven on a single hand, so the actual answer may be between zero and $98 = 7 \times 2 \times 7$, or potentially even more. Now the fraction of people with a number of fingers different from ten is quite low, certainly below one in a thousand, so we can assert with 99% certainty that the answer is 70.[8]

> *Can't we?*
>
> *Think of your response and jot it down.*

Our level of confidence would be accurate as long as the group of people are chosen at random. If the room in question happens to be in a prosthetics laboratory, then the 99% certainty claim might have to be revised. So there are really two quantities to contend with:

- How good is your claim? Or equivalently, within what range do you make it?
- With what certainty do you make your claim?

Very often, the standard in citing statistics is 95% certainty.[9] This often goes without mention, so sometimes newspaper reporters (or their editors) do not understand this and leave out mention of the measure of confidence from their articles. These details should be included, since the common reader is unlikely to know the convention. So in a statistical analysis, we would try to make our claims with

---

[8] If a person is ten-fingered with probability 99.9%, then seven random people will each have ten fingers with probability $(0.999)^7 \approx 99.3\%$. See Appendix 7.

[9] We discuss confidence levels in Appendix 8.6.

95% certainty. Our analysis of the Civil War is historical, not statistical: we have no control over the data we collect, and the data are imperfect at that. In such a situation, even though you may be making a quantitative argument, you may have to make *qualitative* justifications of the numbers you use.

Since we are producing an estimate, we will not produce a single number of Civil War dead (like 734,951), as that almost certainly will *not* be the actual number. So we widen the range until we can make a more plausible claim (something akin to the statistician's 95%). For example, we can say with great confidence (surely more than 95% certainty) that the Civil War death toll was between zero and 31.4 million (the population in 1860, itself an imperfect estimate), but such a wide range is useless. We want to supply the *narrowest* range of numbers about which we can make a confident claim. *That* is how we will attempt to answer this question. Each assumption we make will allow us to produce numbers but then will also introduce uncertainties that we will try to address.

To illustrate, let us consider Hacker's *assumption* that "the native-born white population of the United States in the late-nineteenth century was closed to migration." Of course, this is not literally true, as people came and went, so he must justify employing this assumption. Here's how he does this. He cites data showing about 56,000 US-born whites residing outside the US in 1851. Now if *all* of these people had suddenly moved there from the US in 1850, then they would be counted falsely as "deaths" by his census methods, since they would fail to appear in the census of 1860. That would seem to be a lot of false deaths, but in fact it represents less than 3% of the measured deaths in the US among US-born whites in the 1850s – so the assumption could lead to a potential error of just about 3% in the final figure. (Granted, the decade of interest is the 1860s, not the 1850s, but the point can be made in this previous decade, for which there are good data.) Yet even that 3% is extremely unlikely, since surely many of those 56,000 were already living in Canada, say, and did not recently move there. Plus, there is a factor that helps offset the error from false deaths from the two-census methods due to people who left the country – specifically, the "false births" from people who immigrated to the US in that decade. So while there was surely some migration, it wouldn't have made a large net impact on the analysis.

Hacker identifies another source of error: census undercount. No census is perfect. Some people are not counted; some are counted twice; so there is a net miscount, which we will call an undercount. (If the census *over*counts – and this is rare – we'll say that the undercount is negative.)

> *If N people are not counted and D are counted twice, what is the net undercount? Think of your response and jot it down.*

> *Answer: N − D. (One double-counted person negates the effect of one uncounted.)*

Precisely how does undercounting affect a two-census estimate? To see, consider trying to count wild kindergartners in a classroom that has 20 students on Day 1 and 24 students on Day 2. If your "census" only counts half of them on Day 1 (so, 10) but catches all of them on Day 2 (so, 24), then you will incorrectly conclude that 14 new students had been added between the days, rather than the correct answer of 4. However, if both censuses undercounted the students by precisely 5 each day (maybe the same 5 hid under their desks), then you would have counted 15 students on Day 1 and 19 students on Day 2, correctly concluding that 4 students were added. So the source of error is not undercounting but the *difference* of the undercounting between the two censuses.

By now, you may have anticipated the general scheme: try to get a handle on the undercounting, then adjust your computation accordingly. Here's an easier question to consider: how many sheep in my pasture? Counting livestock is an important business, considering that, for example, to the farmers and ranchers who raise them, animals may be worth hundreds or even thousands of dollars each. When the herd is piling in to feed, or get shorn, or milked, or whatever, they don't sit still and wait. In the old days, people had to count them, and people made mistakes. (Even modern scanners make mistakes, as you may know from supermarket checkouts.) Suppose you do a controlled experiment by having an experienced cattle rancher take his or her time counting heads in a pen, then see how much a rookie youngin' counts as the herd files past. Do this a number of times, and you can calculate the average error someone makes. You may be able to see that a sheepcounter is likely to have a fixed error *rate*; that is, he or she may be off by a constant *fraction* of the total number of sheep, e.g., missing three out of every hundred (3%) in the flock.

Returning to our classroom, we can model our undercount as a percentage error by supposing that one-fourth of the children hide under the desks. Then our undercounting will consistently be 25%. So we will count 15 for the first day and 18 for the second day, concluding that 3 children were added. But if we *knew* that we were undercounting by 25%, then we could conclude that our 3 was too small by 25% and that the actual answer is 4.

Hacker's estimate includes an assumption that the undercounting of the US Census at the time lay between 3.7% and 6.9%, with a preferred value of 6.0% – and we will only use this latter figure, for simplicity. (Keeping the high and low numbers would produce a range of estimates, as discussed above.)

> Suppose we find a two-census difference of d but know that there is a systematic 6% undercounting error in each census. What should we use for the actual difference?
>
> Think of your response and jot it down.

Calling D the actual difference, we know that $d = D - 0.06D$, so $D = d/0.94$.

Hacker is meticulous about his assumptions. We will state only some of them here, as we ultimately shall build a slightly simpler model for pedagogical purposes.

Assumptions

- The native-born white population of the US in the late nineteenth century was closed to migration. Justification: based on estimates of US citizens living in the most common ex-pat country, Canada, even in the unlikely event that *all* of them had moved there between censuses, the effect would only have been a few percentage points.

- Changes in the net undercount of the native-born white population among the four censuses affected males and females equally. Justification: an academic study of census undercount suggests this, as the estimated native-born male and female undercounts show a strong correlation.

- War-related mortality among white females age 10–44 was negligible relative to war-related mortality among white males age 10–44. Justification: there were certainly war-related deaths due to food and supply shortages as well as some direct targeting of civilian populations, each of which could have caused female war-related deaths. (There were also female military deaths – see immediately below this list.) Hacker cites an estimate of 50,000 civilian war-related deaths and shows that if this was somewhat close, the number of war deaths of white females within this age range (historically a hardy lot compared to other demographics) would have been just about 9,000, which would not have affected the analysis much (9,000 is just over 1% of 750,000).

- The expected normal age pattern in the sex differential in survival for the 1860s is best approximated by averaging the sex differentials in survival observed in the 1850–60 and 1870–80 intercensal periods. Justification: this assumption is the same as saying that the graph of the survival rates would have been approximately linear in the time near the war (see Appendix 6.2), an assumption that should be valid if there are no great shifts in survival rates between neighboring decades – and none have been found.

- Foreign-born white males experienced the same rate of excess mortality due to the war as native-born white males. Justification: we have academic studies of the two demographics in aggregate and within a single company.

- The net census undercount of white men aged 10–44 in the 1860 Census was between 3.7 and 6.9%, with a preferred estimate of 6.0%.

- Thirty-six thousand black men died in the war. Justification: this is the official estimate of the Union army, and it is preferable to the two-census method due to the expected high rate of *civilian* black deaths relative to military black deaths, for several reasons (high mortality associated to migration and the transition from slave to free labor, as well as postwar violence).

- Excess male mortality in the 1860s was due entirely to the American Civil War. Justification: though there may have been other significant causes of death within the decade (such as disease), there is little doubt from the historical record that war was by far the dominant cause.

We would be remiss not to point out that there were women who fought in the guise of men during the Civil War and in many other wars throughout history when they were not permitted a combat military role. The contributions of such brave women should neither go unnoticed nor uncounted. However, this secretive group is a very difficult demographic to count, and there is good reason to believe that the total number of female soldier deaths would not be enough to affect our analysis. A conservative estimate of female Civil War soldiers has placed the number in the hundreds.[10] Even a 100% mortality rate among such soldiers would not, as we shall see, affect our total count, given the degree of precision we are able to employ. (Nonmilitary female war deaths were addressed in the third assumption above.)

Note that while the assumptions allow one to control which variables will contribute to the mathematical model – enabling the analysis to be "self-contained" – the justifications themselves rely on citations of previous works. (In Hacker's article, some comments about the methods and reliability of those works are made, often with regard to the application of statistics in arriving at conclusions.) This is par for the course when you ask real-life, difficult questions: you can't fit the whole analysis under one roof.

Building the Model

Suppose for simplicity that the census shows just 1,000 men in their 30s in 1860 and 800 men in their 40s in 1870 (they have aged ten years). What would we conclude? The assumption is that no men entered or left the country during this time, so we would deduce that 200 of the men had died.

> *Could we attribute these deaths to the war?*
>
> *Think of your response and jot it down.*

No. Men would have died for other reasons, too. We have to compare this number to the number who would have died *had there been no war*. Essentially, this number is obtained by computing the fraction of men in their 30s in 1850 who died in the 1850s and the fraction of men in their 30s in 1870 who died in the 1870s, and taking the *average* of these numbers: this average gives our estimated fraction of men in their 30s in 1860 who *would have died* in the 1860s had there been no war. We can then attribute the difference between this number and the 200 who actually died to wartime deaths. That's it. We then do the same with other cohorts.

The rest of the complication of the actual model is around trying to adjust for errors due to imperfect census data. Hacker's model is slightly more complicated than the simplified one that we will employ for pedagogical purposes. Here's our model, in a nutshell:

1 Get the survival rates in the 1850s by finding the percentage of the native-born white male population in a particular cohort that lived from 1850 to 1860.
2 Repeat this for the same cohort of men in the 1870s.
3 Average the two rates above and call the result p. This is the assumed peacetime survival rate for native-born white males in that cohort for the 1860s.
4 Repeat Step 1 for the 1860s and call the result w. This is the wartime survival rate for native-born white males in that cohort in the 1860s.

---

[10] www.civilwar.org/education/history/untold-stories/female-soldiers-in-the-civil.html.

5 Subtract the wartime survival rate for native-born white males from the peace-time survival rate p: this quantity $r = p - w$ represents the fraction of native-born white males dying due to the war – but we cannot simply use this, due to census undercounting.

6 Divide the census measure of white males (not necessarily native born) in this cohort in 1860 by 0.94 to account for census undercounting and get the "true" white male population in this cohort in 1860.

7 Multiply this "true" population of white males in this cohort in 1860 by r. This is an estimation of white male deaths (including the foreign born) due to the war in this cohort.

8 Sum up the above results for each male cohort between 10 and 44 years of age.

9 Add 36,000 to the result to account for deaths of black men.

Example 2.1 Let's try to implement some of Hacker's algorithm in action for a particular cohort. Here is a table of some census counts for native-born white males aged 25–29 years and 35–39 years.

Year	25–29	35–39
1850	654,370	
1860	855,794	584,639
1870	950,049	692,199
1880		920,264

The 1850s survival rate for native-born white males between 25 and 29 is $584,639/654,370 \approx 0.89344$. Then for the 1870s it is $920,264/950,049 \approx 0.96865$. The average of these rates is $p = 0.93105$, the assumed peacetime survival rate for the 1860s for this cohort. The actual 1860s survival rate for this cohort was $w = 0.80884$, and the difference $r = 0.12221$ is the death rate due to the war, a rather large (unadjusted) figure of about 12% of men in this cohort. We have completed Steps 1–5. Now dividing 855,794 by 0.94 and multiplying by r gives a figure of 111,262 white male war deaths in this cohort. We have completed Step 7. ▲

Hacker performs the above analysis for each cohort (Step 8) with low-end estimates and high-end estimates of census undercounting. Adding 36,000 (Step 9) to this figure enables him to produce a range of numbers (between 630,000 and 870,000) inside of which the actual figure of Civil War dead most probably lies, with the preferred estimate of about 750,000.

Conclusions

With the number in hand, we can ask if it makes any sense at all. Being somewhat comparable to existing numbers – but higher – the figure does not seem out of the ballpark. As with all mathematical models, some of the assumptions used in creating it can be challenged or refined, but we have arrived at a figure that helps us to examine the past, and a method that can be employed in other contexts.

Using historical records and analytical savvy, Hacker uncovered a *gap* in the casualty estimate more than twice the human toll of US soldiers in Vietnam – a figure even more striking when computed as a percentage of the population. Recall that the US death toll in Vietnam represented 0.03% of the population. The *difference alone* of 90,000 between the old estimate and the new estimate of deaths in the Civil War represents about 0.3% of the population back then: ten times the percentage toll from Vietnam! This remarkable, meaningful result is a testament to the power and importance of quantitative reasoning.

Summary

We asked how many people died in the Civil War. Before proceeding, we gained an understanding of why the question was important. We then carefully defined what we meant by the question: deaths *attributable* to the war. Since no complete roster of the dead exists, we needed to make assumptions to carry us from the data we had to the figures we need. Hacker is very careful in outlining his assumptions, so this chapter serves as a model for this stage of analysis, at least qualitatively. Assumptions in hand, we built a two-census model for tallying war deaths: find the expected survival rates for the wartime period and compare with the observed number in that decade; the differential is deemed to represent wartime deaths. We summed up these numbers for all cohorts actively involved in the war. After tallying the numbers, we obtained a death toll based on the preferred census undercount estimate. We concluded that the prior "accepted wisdom" was *off* by huge amount, itself a fraction of the population ten times the fraction who died in Vietnam.

Exercises

1 Here are some data on intercensal survivorship probabilities by age, from Hacker's article (the age is determined at the first census):

Age	Males			Females		
	1850–60	1860–70	1870–80	1850–60	1860–70	1870–80
10–14	0.9203	0.8768	0.9780	0.9674	0.9694	0.9987
15–19	0.8946	0.7699	0.9606	0.8178	0.8192	0.8588

a What is the sex differential in survivorship from 1860 to 1870 at aged 10–14, i.e., the difference (as a percentage) between males and females in this cohort?

b If the population of males aged 10–14 in 1850 was 1,147,038, what can you say about the population of males aged 10–14 in 1860? What about the population of males aged 20–24 in 1860? What if we take into account our estimate of a 6.0% undercount?

   c Can you calculate the excess mortality rate of males aged 15–19 due to the war?

   d Can you calculate the intercensal survivorship probabilities of females aged 10–19 from 1860 to 1870? If you can, do so. If not, which data are missing?

   e The intercensal survivorship probability for females aged 5–9 from 1870 to 1880 was 1.0236. How might you explain this result?

2 Pick any stock on any exchange and track its closing price for five consecutive weeks one year and the same five consecutive weeks five years earlier (the stock will need to have been listed for at least five years). Average the percentage increase over the five weeks. Compare how this average has changed over five years. (This exercise involves research.)

3 The table below provides data on migration between the US and Canada in 2012 and 2013.[11] "Permanent Residents" refers to the number of new permanent residents in the specified country who came from the neighboring country. Assume that the new permanent residents are included in the total population of their new country only.

		2012	2013
Population	US	314,112,078	316,497,531
	Canada	34,754,312	35,158,304
Permanent Residents	US	20,138	20,489
	Canada	7,891	8,495

   a What fraction of Canadians immigrated to America in 2012?
   b What fraction of Americans immigrated to Canada in 2013?
   c Assuming the choice to move across the border is random, what fraction of Canadians moved to the US in 2012, then back to Canada in 2013? How many people is this?

4 Find US Census data to calculate the death rate for women the age of your biological mother (or any parent or guardian) at the time of your birth, in the decade you were born – and do the same for that cohort 100 years earlier. (This exercise involves research.)

5 Here are some data in addition to the data presented in the text. In 1850, there were 452,270 native-born white males aged 35–39. In 1860, there were 400,900 native-born white males aged 45–49, while in 1870, that number

[11] Population data, http://data.worldbank.org/indicator/SP.POP.TOTL?cid=GPD_1. US Permanent Residents from Canada, www.dhs.gov/sites/default/files/publications/ois_yb_2013_0.pdf. Canada Permanent Residents from US, original site now unavailable; most relevant current site is www.statcan.gc.ca/pub/91-209-x/2016001/article/14615/tbl/tbl-03-eng.htm.

was 496,808, and in 1880, it was 621,164. Assuming a census undercount of 6%, estimate the number of war deaths among native-born white males aged 35–39 who died in the 1860s.

6 Think of a topic for a project related to this question.

Projects

A Explain how it is possible to create an imperfect digital "restoration" of a damaged painting and how the process is similar to estimating the would-be deaths of males in the absence of the Civil War. Be detailed and include examples.

B If the US government were to decide to make reparations for slavery in the form of financial restitution, how might it arrive at a dollar figure?

C Conduct a two-census estimate of American male deaths from World War II.

3

How Much Will This Car Cost?

Appendix Skills: Arithmetic, Algebra

OVERVIEW

We study interest on a car loan – why it is charged the way it is, and how it accrues in exponential fashion. To understand this real-world example, we do the following:

1 We go through some of the fine print on a car financing site, setting a goal of understanding a single sentence.

2 We consider the basics of lending and interest.

3 We discuss the process of accruing interest just once, *then what happens when we do this repeatedly.*

4 We derive the function which represents a loan balance over time, including the reductions due to regular payments.

5 We return to the fine print with a finer understanding of the finances of loans, monthly payments, and all that.

Legalese and Gobbledygook

Here is some fine print from a bank website on car financing:[1]

> Automatic payments from a U.S. Bank package required. Rates as low as
> 2.49% APR are available for 3-year auto loans $10,000 and higher at 100%
> loan-to-value (LTV) or less. Rates for loans to purchase a vehicle from a private
> party, smaller loan amounts, longer terms, vehicles older than 6 model years,
> or higher LTV may be higher. Loan fees apply. Loan payment and APR will vary
> based on the loan amount, the term, and any fees. Origination fees vary by state
> and range from $50 to $125 or up to 1% of the loan amount. Loan payment
> example: a $10,000 automobile loan at a 1.68% Interest rate for 36 months with
> a $125 origination fee will have a 2.49% APR and a $288.59 monthly payment.
> Offer is subject to credit qualifications. Rates are subject to change. Some
> additional restrictions may apply. Installment loans are offered through U.S.
> Bank National Association. 2013 U.S. Bank. Member FDIC

---

[1] www.usbank.com/loans-lines/auto-loans/new-car-loan.html. N.B.: specific numbers on the website are changed by the bank over time, though the content seems stable.

Huh? Look, no one reads this stuff. But buyer beware: we ignore this at our peril. We wouldn't pay twenty dollars for a scoop of ice cream, but if we lack the tools to analyze major financial decisions on our own, we might unwittingly squander hundreds or thousands by signing up for a bad loan. Daunting as it may seem, we must understand our financial undertakings.

Paragraphs like the one quoted above are written by highly trained people with advanced degrees. They are written to be precise, not to be user friendly or even particularly clear. Understanding such a paragraph is possible, but you have to be patient and break it down piece by piece (see "Complexity vs. Depth" sidebar below). Let's try to understand this *one* sentence:

> Loan payment example: a $10,000 automobile loan at a 1.68% Interest rate for 36 months with a $125 origination fee will have a 2.49% APR and a $288.59 monthly payment.

Interesting Thoughts

First of all, what is an interest rate? Even that one phrase requires a separate discussion – but bear with us, we'll get there. (Oh, and don't be offended if things seem "too easy" or "too babyish" at first. That's the idea: make the difficult simple.) Suppose you are rolling in money and your sister – or maybe a distant acquaintance, or a complete stranger – wants to borrow $100 for some headphones and says she'll pay you back next month. Would you do it? And what if she says she'd pay you back in a year? Probably, the answer depends on how well you know this person, and how likely you think it is that you'll ever see your money again.

In lending money, you are providing a service – giving her something she wants – and in doing so, you incur some risk: the possibility that she will not pay you back. Of course, you may lend from the goodness of your heart, but if you were in the *business* of lending money, your business would charge a fee for the service that it provides. That fee would be larger, as well, if the risk of getting the money back was greater (e.g., a complete stranger rather than a trusted relative).

Okay, so you're going to charge *something* to provide the service of lending customers money when they want it – money that they don't have. Let's say that you agree that your customer will pay you one dollar (extra) for the loan, so will pay you back a total of $101 in one month's time. So that's the cost: one dollar for a loan. That cost – the money you get back beyond the original amount you lend – is called the *interest*. In this case, the interest you gain after one month of the loan is $1. But what if after you two agree, she decides she needs *two* sets of headphones and wants to borrow $200 dollars?

> *Would the price of that loan still be one dollar?*
> *Think of your response and jot it down.*

In theory, you could just find someone else to loan the second hundred dollars to, and make two dollars from the two separate loans. So shouldn't you charge two dollars for a $200 loan and three dollars for a $300 loan, and $2.75 for a $275 loan? In other words, your fee is a *fixed fraction*, or fixed percentage, of the loan amount. That percentage is called the "interest rate." In this case it is one-hundredth, or 1%. The bigger the loan, the greater the service, the larger the cost.

(Now I'm tempted to write "that was easy," but I'm not sure it was. What *is* true is that that's all there is to it, and it is understandable with some focus. So whatever the case, easy or not, there it is.)

Complexity versus Depth

Completing a thousand-piece jigsaw is a difficult task. So is playing flute in an orchestra. But they are different in nature. The jigsaw is *complex*, while playing music is *deep*. Not to diminish competitive jigsaw enthusiasts, but the basic element of the game – testing to see if one piece fits into another – is simple. What makes the task hard is the sheer number of possibilities. (Of course there *are* strategies for managing the job: go for the borders; separate by color; look for pieces with lettering on them.) Playing flute – or basketball, or chess, or dancing on stage, or writing a book – involves managing an array of interacting goals by drawing from a panoply of individual skills.

Generally, if something difficult does not require years of training or a specialized expertise, it is probably because of complexity rather than depth. This includes almost everything in this book! (An exception might have been the proof of the Central Limit Theorem of Section 8.5, but we omitted that.) That means that you can master it by breaking it down into its component pieces and understanding them. It also means there is no reason for fear!

Let's discuss a potentially tricky point. What if your customer decides that she wants to pay you back after a full *year*, not a month. Then would you still charge one dollar?

What would you propose as a fee?

Think of your response and jot it down.

You may have realized that you would be able to repeat the same loan twelve times in that year, so perhaps your proposed fee was $12. Very often, interest rates are written as "annual interest rates," even if not explicitly stated as such. So in this scenario you would charge a 12% annual interest rate on your one-year loan.

If you (or a bank) decides that the interest paid for a loan that's one-twelfth as long is one-twelfth the interest – and likewise for any fraction of the length of the time period – this is called "simple interest." That is, simple interest describes interest that is *proportional* to the duration of the loan: if you plotted the amount of simple interest you owe as a function of time, the graph would be a straight line through the origin (see Appendix 6.2).

Do you think banks charge simple interest on their loans?

Think of your response and jot it down.

They hardly ever do! Why wouldn't all banks always charge simple interest? The answer is the key to really understanding interest and many other kinds of nonlinear phenomena. "Nonlinear"? As we will see, the graph of interest owed as a function of the time of the loan is *not* a straight line!

Compounding the Problem

So far, we understand why interest is charged (the price of a service; to cover the risk of default), why it is proportional to the loan amount (because we could divide a large loan into many small loans), and why you get more interest in more time (because we could divide a time period into many smaller ones). We now want to consider simple interest versus an alternative.

Okay, now suppose you start out this whole lending campaign with $100 and you lend out all of it, charging 1% on a one-month loan. After one month, you will have $101 dollars. Now suppose you lend all of *that* out to someone you find (say) who wants to borrow exactly $101 dollars at the same rate.

How much interest will you earn in that second month?

Think of your response and jot it down.

Well, you would earn 1% of $101, or $1.01. An extra penny! So now you have $102.01 in cash after two months. That extra penny is the 1% interest on the dollar of interest that you earned from the first loan. By the same reasoning, if your borrower asked to pay you back after two months instead of one, you can think of the second month as a loan of $101 dollars (the amount she owed you after one month), from which you would collect $1.01 in interest that month. You could charge her a total of $2.01 in interest. Interest upon interest is the main reason that the banks' profit is not a linear function of time: $2.01 is more than twice $1.00.

After the second month, we'll have to consider fractions of pennies – or for fun we could imagine starting with a hundred *million* dollars, and these fractions of a million dollars would make sense. So let's just continue, if you'll allow it. Each month, you lend out all the money (possibly by allowing a single borrower not to return it) and earn 1% interest. So after each month, you have 101% of what you started with. Right? Now, a small math step: put differently, this is 1.01 *times* what you started with. In fact, the words tell you this, in code. Recall *percent* means "out of a hundred," so 101% means 101/100, and this is 1.01 in decimal form. Also note that *of* means "times" – half *of* ten means half *times* ten, or five – so 101% *of* the starting amount means 1.01 *times* the starting amount. Stop and think about it.

If you think you've got it, let's try an exercise. Suppose you begin by lending out P dollars. (I'm using a variable because I want to imagine many different situations without repeating myself many times.) Riddle me this:

How much money will you be owed after one month?

Think of your response and jot it down.

The answer is $(1.01)P$. As a reality check, when P was 100, the answer was $(1.01)(100)$ dollars, or $101. Then when P was 101, the answer was $(1.01)101$ dollars, or $102.01. Great. So here's how the steps go, month by month. You can imagine $P = 100$, but any number will do:

$$P \to (1.01)P \to (1.01)(1.01)P \to (1.01)(1.01)(1.01)P \to \cdots$$

Each month, we multiply by 1.01. (Note that the number 1.01 is 1 plus the 0.01 interest rate.) We've worked our way up to an important question:

If we start out with P dollars, how much money will we have 12 months later?

Think of your response and jot it down.

The answer is $(1.01)^{12}P$. Cue the calculator! This is about $1.127P$. Note it is **not** $1.12P$, which was the 12% simple-interest rule that we discussed above. In fact, we have yielded 12.7% interest. This is a central point: the two methods of charging interest earn you different amounts! The interest-upon-interest phenomenon led to an extra 0.7% revenue over the course of the year.

Start with 80. Add to it a quarter of that. Add to the resulting number another quarter. What do you get?

Think of your response and jot it down.

You go from 80, to $80 + 20 = 100$, to $100 + 25 = 125$, adding a grand total of 45. This is 5 more than two quarters (40), the extra 5 being one quarter of the first 20 added: interest upon interest.

APR and APY

We found that even though our interest rate was nominally 12%, we actually paid 12.7% over the course of the year. Curious, but is this really a big deal? In real dollar amounts, what's the difference? We have that $1.12 \times \$100$ is $112, while $1.127 \times \$100$ is $112.70, a difference of just 70 cents! (Maybe now is the time to imagine that we are lending a hundred million dollars, for then the difference would be 0.7 million, or $700,000.)

But even for small amounts, these differences add up. Inching closer to our main example, let us suppose you lent out $10,000 for ten years at an interest rate of 12% for a year. Then after 12 years at *simple* interest you would earn ten times your yearly interest of $1,200 (this is 12% of the loan amount), or a total of $12,000. (This interest is the amount owed to you on top of the actual loan amount, or *principal*, so the payment would actually be $22,000.) But if you did not charge simple interest but instead the amount that you would make by repeating the process of re-loaning everything each year at 12%, then after ten years you would collect $(1.12)^{10}$ times the loan amount of \$10,000. Cue the calculator! $(1.12)^{10}$ is about 3.106, so you collect $3.106 \times \$10,000$, or about \$31,060. You earn \$21,060 in interest, and the difference from your earnings at a simple rate is a whopping \$9,060! Even better, if we return to charging 1% interest each *month* for ten years (120 months), we get $(1.01)^{120}$ times \$10,000, which turns out to be about \$33,000, an *extra* \$11,000 beyond simple interest. It pays – literally – to know about how interest piles up when you recalculate the loan balance at each time interval instead of using simple interest. The process of calculating interest at fixed intervals is called *compounding*.

Unfortunately, most of us, more often than not, are on the borrowing end of this equation. This means more interest that we have to *pay* to the banks who have lent us money for our homes, cars or other major life events – even more reason to be vigilant about the fine print.

The banks must tell you how often they calculate interest: quarterly, monthly, weekly, daily, every second ... nanosecond ... *continuously?* This information will be part of your contract. The sooner the process of interest-upon-interest begins, the more interest will accrue. So the banks want to calculate/compound interest more frequently to make more money. Market pressures, conventions and expectations depending on the type of loan also play a role in determining what the bank decides.

Let's assume as in our main example that you borrow \$10,000 at an annual interest rate of 1.68%, compounded monthly.

> *What percentage do you owe in interest each month?*
>
> *Think of your response and jot it down.*

Since one month is one-twelfth of a year, the answer is one-twelfth of 1.68%, or 0.14%, also known as 0.0014. So at the end of the month, you owe 100% plus an extra 0.14% of what you owed at the start, or 1.0014 times that amount. If you owed \$10,000 at the start, then after one month you owe $1.0014 \times \$10,000 = \$10,014$. The amount you now owe the bank has grown by a factor of 1.0014.

> *After 12 months, by what factor has the amount you owe grown?*
>
> *Think of your response and jot it down.*

The answer is $(1.0014)^{12}$, or about 1.0169. This means the annual percentage *yield* (APY) – the actual percentage of interest collected after one year – would be 1.69%, slightly higher than the annual percentage *rate* (APR) of 1.68% used to determine the amount collected in each compounding period.

> *Calculate the annual percentage yield of a 6% loan compounded daily.*

To answer this, we note that each day, we earn interest of $\frac{1}{365} \times 6\%$, or $0.06/365 \approx 0.00016438$. So each day, our loan grows by a factor of 1.00016438.[2] After 365 days

---

[2] Okay, this is crazy, but banks often use a 360-day "year": 30 days in a "month," 12 "months" in a "year." Since our main example involves compounding monthly – and we all agree there are 12 months in a year – I won't add this complication.

it grows by $(1.00016438)^{365} \approx 1.0618$, giving an annual percentage rate of 6.18%. It's counterintuitive, perhaps, but if you take out a loan at 6% (APR), then after one year, you actually pay an interest rate (APY) of 6.18%

What if the interest rate (APR) was r, thought of as a fraction (not a percentage)? Then after one day we would owe $r/365$ in interest, so our loan balance would be $1 + r/365$ times what it was at the start of the day. This calculation is true for each day, so if we start by borrowing a principal amount of P dollars, the sequence goes

$$P \to (1 + r/365)P \to (1 + r/365)(1 + r/365)P \to \cdots,$$

and after 365 days we owe $P(1+r/365)^{365}$. After t years, we owe $P(1+r/365)^{365 \cdot t}$. This is an exponential function of the form ab^t, where $a = P$ and $b = (1+r/365)^{365}$. Eventually, it becomes large and steep – see Appendix 6.4 – a mathematical image for what being "trapped under a mountain of debt" can mean to longtime borrowers who can't keep up with interest payments.

Remark 3.1 At this point, we are also able to figure out what would happen if we compounded interest *continuously*. We first note that if we compound it n times in a year, then a loan of P increases by a factor of $(1 + r/n)^n$. When n gets large and approaches infinity, this quantity[3] turns into e^r. Then after t years, the loan balance would go

$$P \to Pe^r \to P(e^r)^2 = Pe^{2r} \to Pe^{3r} \to \cdots \to Pe^{rt}.$$

This formula is sometimes called "pert" and is good for population growth or radioactive decay or other processes that change *continuously* in proportion to their amounts.

There is another perspective on this exponential function e^{rt} that will be helpful. In a small amount of time Δt, the loan balance changes by a fraction $r\Delta t$ of itself (the above example of daily compounding being the case $\Delta t = 1/365$), or in other words, by an absolute amount $Pr\Delta t$. Such a property is a hallmark of the exponential function: its rate of change is proportional to its value, and here r is just the constant of proportionality. It happens that then the change in value after any fixed amount of time is always a fixed amount – see Appendix 6.4.   ▲

Returning now to the case of interest compounded daily, we can ask, what extra fraction have we actually paid in a year? This extra fraction is what we called the *annual percentage yield*, or APY. If the APY were R (not little r, big R), then this means that we would pay the fraction R of P (or RP) in interest, meaning the total should represent $P + RP$, or $(1+R)P$. Comparing $(1 + r/365)^{365}P$ to $(1+R)P$, we can divide out P and find $1 + R = (1 + r/365)^{365}$, or

$$R = (1 + r/365)^{365} - 1$$

---

[3] You will either have to take this statement on my authority or use calculus, specifically L'Hôpital's rule.

so

$$APY = (1 + APR/365)^{365} - 1,$$

where we have recalled that r was the annual percentage rate *APR*. In this equation, 365 is the number of times we compound. For the car loan we would use 12. If we compound n times, then APY $= (1 + APR/n)^n - 1$. (Use $n = 360$ for a banker's year.)

You can now make your own APY calculator!

> *What is the APY on an interest rate of 18%, compounded daily?*
>
> *Think of your response and jot it down.*

As a warning to credit card owners, 18% translates into a 19.7% APY, almost 2% higher.

Monthly Payments

The Rule of 70

One rule of thumb is that if a population (such as the population of dollars that you owe on a loan) grows by x% a year (or whatever unit of time), then it will double in 70 divided by x years (or whatever unit of time). So, for example, growth at 10% per year should double in about $70/10 = 7$ years. An ant colony that grows at 5% per week will double in 14 weeks.

To see why, consider that a population growing by 10% a year will increase by a factor of 1.1 annually and so will double in T years if $1.1^T = 2$. To find T, take the natural logarithm of both sides: $T \ln(1.1) = \ln(2)$. We find $T = \ln(2)/\ln(1.1)$. Cue the calculator! We get $T = 7.3$.

More generally, growth by x% means the yearly growth factor is $1 + x/100$, so by a similar argument, we find $T = \ln(2)/\ln(1 + x/100)$. What makes this work is that the logarithm of $1+$ *something small* is approximately equal to that *something*, so $\ln(1 + x/100) \approx x/100$, and we find $T \approx 100 \ln(2)/x$, or $T \approx 69.3/x$.

Returning to the example of the chapter, we found that an APR of 1.68% compounded monthly led to an APY of 1.69% – but the US Bank website quoted at the start of this chapter talks about a 2.49% APR. What's the deal!? Where did this come from? Reading the text more carefully, we see that it says there is a loan origination fee of $125. How does that factor into the calculation?

> *Any ideas?*
>
> *Think of your response and jot it down.*

Since the example talks about "monthly payments," we will assume the plan is that we will be paying a fixed amount (evidently $288.59) each month until we owe nothing. And part of what we owe is the $125 fee. So let's tack that onto the $10,000, meaning at the start we owe $10,125. Therefore, we will try to answer the question,

> *What fixed amount would we have to pay each month to pay this off in 36 months?*

Next, we will explain the mysterious APR by asking,

*If there were **no** loan origination fee, what APR would lead to this monthly payment?*

Let's calculate the first question first. Since we plan to pay back some amount each month, our previous calculations must be performed anew – they didn't take into account that the amount owed would grow with interest but also *decrease* by the amount of the monthly payment.

So suppose we pay back M dollars each month. Lots of letters now. P is the amount we borrow (principal); r is the annual interest rate (APR) expressed as a fraction, not as a percentage; M is the amount we pay back each month. After one month we owe

$$P + \frac{r}{12}P - M,$$

the loan amount P plus interest $\frac{r}{12}P$ minus the payment M. Reality check: it better be the case that M is larger than $\frac{r}{12}P$, or else we owe more at the end of the month than we did at the start, and we're never going to pay off the loan! (The difference between M and rP is what reduces the principal, the amount of debt remaining from the original loan, not just the interest on it.) We can write this amount as

$$\left(1 + \frac{r}{12}\right)P - M.$$

A similar process will be repeated each month: the balance increases by a factor of $1 + \frac{r}{12}$ but decreases by an *amount* (not a factor) M. Let's make a table.

Months since loan	Loan balance
0	P
1	$P + \frac{r}{12}P - M = \left(1 + \frac{r}{12}\right)P - M$
2	$\left(1 + \frac{r}{12}\right)\left[\left(1 + \frac{r}{12}\right)P - M\right] - M$
	$= \left(1 + \frac{r}{12}\right)^2 P - \left[1 + \left(1 + \frac{r}{12}\right)\right]M$
3	$\left(1 + \frac{r}{12}\right)^3 P - \left[1 + \left(1 + \frac{r}{12}\right) + \left(1 + \frac{r}{12}\right)^2\right]M$
$\vdots$	$\vdots$
N	$\left(1 + \frac{r}{12}\right)^N P - \frac{\left(1 + \frac{r}{12}\right)^N - 1}{\frac{r}{12}}M$

Simplifying the large term in brackets at the end involves a little trickery,[4] but ultimately we learn what we owe after N months. If we want the loan to be paid

---

[4] First verify the following identity: $(1 + x + x^2 + x^3)(x - 1) = x^4 - 1$. You have to distribute and cancel a lot of like terms. Dividing by $x - 1$ gives $1 + x + x^2 + x^3 = \frac{x^4 - 1}{x - 1}$. Now if x is set equal to $1 + \frac{r}{12}$, we find $1 + \left(1 + \frac{r}{12}\right) + \left(1 + \frac{r}{12}\right)^2 + \left(1 + \frac{r}{12}\right)^3 = \frac{\left(1 + \frac{r}{12}\right)^4 - 1}{\frac{r}{12}}$. The same trick works with similar sums, so $1 + \left(1 + \frac{r}{12}\right) + \left(1 + \frac{r}{12}\right)^2 + \cdots + \left(1 + \frac{r}{12}\right)^{N-1} = \frac{\left(1 + \frac{r}{12}\right)^N - 1}{\frac{r}{12}}$.

off in N months, we set the quantity at the bottom equal to *zero*. This gives an equation for us to solve to find the monthly payment, M.

Solve for M as described.

Think of your response and jot it down.

A little algebra gives $M = \frac{\frac{r}{12}(1+\frac{r}{12})^N P}{(1+\frac{r}{12})^N - 1}$. You can divide numerator and denominator by $(1+\frac{r}{12})^N$ to get

$$M = \frac{\frac{r}{12}P}{1 - \left(1+\frac{r}{12}\right)^{-N}}.$$

We've just figured out how to create a mortgage calculator! That is, we determined the regular payments M required to pay off a loan amount P with interest rate r after N periods of time.

With this information in hand, we can return to the question at the start of the problem. If the loan amount is \$10,125 (the loan plus origination fee) and we pay an interest of $\frac{r}{12} = \frac{1.68\%}{12} = 0.14\% = 0.0014$ each month, and if we want to pay this off in 36 months, then we find that

$$M = \frac{0.0014 \times \$10,125}{1 - 1.0014^{-36}} \approx \$288.59$$

This is the amount cited in the loan example, which we recall here:

Loan payment example: a \$10,000 automobile loan at a 1.68% Interest rate for 36 months with a \$125 origination fee will have a 2.49% APR and a \$288.59 monthly payment.

The monthly payment corresponds to a 36-month loan in the amount of \$10,000 *plus* the \$125 loan origination fee.

Now we come to the second question: if there were no loan origination fee, what interest rate would lead to the same monthly payment of \$288.59? To answer, we use the same equation

$$\$288.59 = \frac{\$10,000\frac{r}{12}}{1 - \left(1+\frac{r}{12}\right)^{-36}},$$

but this time we have used the known $M = \$288.59$ and instead solve it for r. We find $r \approx 0.0249$. So with no fee, the APR would have to be 2.49% to lead to the same monthly payment. By sneaking the fee in – even though you never see it because it gets rolled into the principal amount of the loan – the bank effectively gets to charge you a higher rate.

This solves the question entirely; we have completely understood one sentence!

Summary

To summarize, we took a single sentence from an auto loan example provided by US Bank and tried to figure out what it meant.

We learned that we pay interest on loans as payment for the convenience and to compensate the lender for the risk of default. By considering breaking up larger loans into smaller pieces, we figured out why interest is proportional to the loan balance, adding a fixed fraction each period. Over time, interest is therefore charged against not only the original amount borrowed (the principal) but also on top of any previous interest that had accrued. Interest on interest can add up significantly over time, reflected in the eventual steep slope of the graph of an exponential function.

Mathematically, the key was basically this: to add a fraction f of an amount A to itself, we take A and add fA. This equals $A + fA$, or $(1 + f)A$. In the example, f is the APR divided by the number of times we will compound in a year. The point is that we begin with some amount and multiply by this same factor $(1 + f)$ many times – i.e., each time we compound – leading to an exponential function of time.

With a quantitative knowledge of how to apply interest to a loan, we could understand the fine print in our example. Specifically, by taking into account monthly payments in our model, we could calculate exactly how incorporating the loan fee into the principal produced an effectively higher interest rate.

Exercises

1 Verify the equation $1 + x + x^2 + \cdots + x^{N-1} = \frac{1-x^N}{1-x}$.

2 You take out a 30-year mortgage for \$200,000 at 5% APR. Assume interest is compounded monthly. What are your monthly payments? Over the course of the loan, how much interest have you paid? How much principal? What is the annual percentage yield (ignoring monthly payments)?

3 Find an actual furniture rental company and compute the true cost of renting a large item of furniture for four years versus paying for it outright. Use actual figures. (This exercise involves research.)

4 Consider the same problem as Exercise 2, except with a 5.4% APR. If one bank was offering 5.4% with a \$2,000 loan origination fee and another bank was offering 6% with no fee, which would you take, and why? Be sure to explain your reasoning and how you performed your calculation.

5 Suppose you plan a trip with friends and are able to purchase transportation and accommodations on your credit card.

Using actual figures (pick a specific place; find real prices for trains, planes, and motels; and locate the terms from an actual credit card), calculate the total cost of paying your credit card bill immediately versus paying the minimum amount every month for a year before settling the bill in full. For simplicity, you may assume no other credit card purchases. (This exercise involves research.)

6 You have retired with a house, a car, and a stash of money but no source of income. We are assuming you live on a self-sufficient farm with solar heating, so that you also have no expenses beyond the following. A security-and-insurance company has offered to guard your assets, guaranteed. It charges you a daily fee calculated from an annual percentage rate of 1% of your assets, which you will pay from your stash of cash. (You may assume a 365-day year. On Leap Day, the company gives you a freebie.) You have $1 million in assets, all told. You plan to live another 20 years. If this is true, what will be the value of your estate at the end?

7 Think of a topic for a project related to this question.

Projects

A When did this dinosaur die? Learn carbon dating, describe the mathematical model for it, and explain an example. Be clear in your explanations, and be sure to discuss several examples and use *different* numbers from whatever sources you employ.

B You're dead-set on buying a luxurious house. The owners are asking $2 million. How are you going to get it done? Be sure to discuss the minimum you will need to earn, your salary over time (see Table 1.1), various mortgage structures (adjusted rate, fixed rate, etc.), and any other relevant factors (realtor fees, property taxes, etc.).

C Should you trade in your gas guzzler for an energy-efficient electric vehicle? Be sure to consider different ways to answer the question, taking into consideration as many factors as you can for each way.

4

Should We Worry about Arsenic in Rice?

Appendix Skills: Units, Probability

OVERVIEW

We ask about a particular personal-health choice, then do the needed work – research, risk estimation, and careful treatment of units – to find an answer that works for us:

1 *First, we look to understand the qualitative features of the questions, what is arsenic, and what are the concerns?*
2 *We study the data and spend some time understanding what research lies behind them and what the numbers mean.*
3 *After learning what is known of risks involved in consuming arsenic beyond a threshold level, we have to figure out how much we consume through our rice-eating habits.*
4 *We do the calculation and make a personal determination that we won't concern ourselves about it.*
5 *More importantly, we recognize that the same steps can be applied to pretty much any such lifestyle decision.*

Sensationalized Risks and Lifestyle Choices

We constantly hear about dangers of chemicals in our food. Eek! Sodium! Caffeine! GMOs! BHA! H_2O! Now H_2O is just water, so not *all* chemicals are bad. Which are okay? Which should we avoid? How to make an informed decision?

As an illustration, we will consider one such question:

Should we worry about arsenic in rice?

What do you think?

Think of your response and jot it down.

It's important to check your gut opinion. You may have just come from a natural foods store scattered with scary literature about the toxins in our foods. Or maybe you just drove home from a visit to your aunt's pristine rice farm in South Carolina, and think there couldn't possibly be any danger in what Farmer Ruth was producing. Whatever your opinions may be, let's recognize that unless you have done a careful study, they *may* simply be opinions with no basis in fact. We'll have to see ...

As with all questions we consider, we should be clear about how we are interpreting it, and what an answer might look like. Here the "we" of this question is not some national association of rice growers but rather an individual consumer who eats rice. An answer to this question will be a decision about whether to change our eating habits, and if so, how.

Our route in addressing this question will take us across varied terrain, including chemistry, medicine, and mathematics. At the end, despite learning a lot about the risks and different choices, we still won't be able to make a conclusive recommendation for everyone, but we can make a choice that is right for ourselves. Now if doing so requires so much work, you can imagine how hard it is for the average consumer to make informed decisions, especially a consumer without quantitative skills! Such consumers usually look for advice in answering these kinds of questions. The best "meta" advice to give them is listen to someone who (1) has some credentials, and (2) has no vested interest in the outcome. For example, clickbait on your newsfeed won't be scientifically researched, while literature from the rice growers' association or a magazine called *Low-Carb Living* won't be unbiased.

So, we will try to determine how much arsenic we are eating in rice and whether this amount should be thought of as dangerous. Simply,

> How much arsenic are we eating in our rice? Is that bad for us?

That's a natural question, but it leapfrogs over something way more basic: what *is* arsenic, and how is it even dangerous?

Wise Up!

We've got to school ourselves on arsenic. I did the work for you.[1] Here's a primer:

- Arsenic is an element, meaning a basic constituent of chemical compounds. There are about 90 naturally occurring elements on Earth, of which arsenic is one. Elements are listed in the Periodic Table, numbered by how many protons appear in the nucleus of one atom. Hydrogen (symbol H) is number 1, oxygen (O) is 8, silver (Ag) is 47, and gold (Au) is 79. Arsenic (As) is 33.
- Different combinations of elements interact differently with one another, leading to vastly different properties and effects on human and animal bodies. For example, sodium (Na) is a metal that will react quite violently with water,[2] producing heat and hydrogen, which is very flammable. This behavior is in striking contrast with familiar metals like iron or copper. Yet a combination of sodium (Na) and chlorine (Cl) produces ordinary table salt (NaCl). Lesson: the properties of the

---

[1] Gathering materials in researching a problem can be a lot of fun. Use whatever media you like, but make note of your sources, and be sure they're reputable!

[2] Take a look: www.youtube.com/watch?v=dmcfsEEogxs.

individual elements in chemical compounds may have nothing to do with the behavior of the compound as a whole.

- Arsenic appears in two main forms, called "organic" and "inorganic." These labels have to do with whether the compounds containing arsenic also contain carbon and have nothing to do with whether the arsenic in the rice came from an "organic" farm. Inorganic arsenic is a known carcinogen, leading to increased risk of cancers of the liver, lung, and bladder, among others. Organic arsenic compounds, such as those that appear in some seafood, are not known to be linked to cancer and are generally considered much less toxic. According to the Integrated Risk Information System of the Environmental Protection Agency (EPA), the studies linking arsenic to cancer come from analyses of inorganic arsenic in drinking water.[3] There are villages in southwestern Taiwan where the well water contains elevated levels of arsenic, and the standard mortality rates (SMR) from various cancers are abnormally high as a percentage of the rest of the population (which has SMR set to 100). Specifically, in the graph below, the ratio of mortality rates of Taiwanese from these villages divided by the mortality rates of Southwest Taiwan are plotted, written as percentages – so a value of 200 means *twice* the chance (against the greater region) of dying from those illnesses.[4]

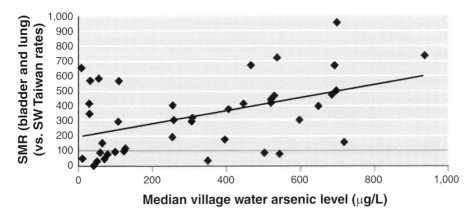

We should spend some time on this graph. We'll pick *one* data point, say, the diamond at height 300 above the *x*-value of 600. To get this data point, whatever it means, you can imagine that a team of researchers had to go to a village water well, take data – probably many times over some time period – and

---

[3] A remark to the reader: because I was not sure that a report linked from a website called Greenerchoices.org would be unbiased, I traced the data to a stable source, the EPA. Stable, but we note that Consumer Reports (http://greenerchoices.org/wp-content/uploads/2016/08/CR_FSASC_Arsenic_Analysis_Nov2014.pdf) cited that the EPA itself failed to report a much higher level of risk than it had earlier indicated in a draft report, due to "political pressure." Absolute certainty is elusive.

[4] Lamm et al., *Arsenic cancer risk confounder in Southwest Taiwan data set,* Environ. Health Perspect. 114 (2006) 10771082, www.ncbi.nlm.nih.gov/pmc/articles/PMC1513326/.

analyze them to determine the median arsenic level. (This comes at considerable time and effort and cost.) Then they would need to know the mortality rates of the village, relative to the region. Those data require a network of hospitals and coroners that take reliable statistics about causes of death, which itself requires considerable health care infrastructure and expense. Scientific knowledge is hard-won; that's what makes it so valuable. After all this data collection – and we have to repeat the whole procedure for *every* point on the graph – the graph seems to reveal a tendency toward elevated cancer risk with more polluted wells (correlation is discussed in Appendix 8.7). At the least, almost every highly contaminated well leads to an SMR above 100. That said, our concern here is about arsenic, not in water, but in *food*, specifically, rice.

- According to the American Cancer Society, inorganic arsenic is used as a wood preservative, in pesticides, in manufacturing some glass and semiconductors, and as an additive to lead and copper. It has not been manufactured in the United States since 1985.

- The arsenic in rice comes from the absorption of water in the soil, and rice is often grown in patties, which assist the absorption of arsenic. Rice grown in different regions can therefore have different levels of arsenic. Also, different rices absorb different amounts of arsenic. Brown rice contains more of the outer parts of the rice grain, where arsenic accumulates, so brown rice of any variety contains a higher amount of arsenic than white rice of the same variety.

- When you measure the amount of one substance (such as a contaminant) in another (such as water), you speak of *concentration*. One way or another (to be clarified below), the concentration controls the fraction of the total represented by the contaminant.

We now have an understanding of the problems associated to arsenic. Now we need to know how they affect us personally, and this will depend on how much rice we eat and how much arsenic it contains. We'll need to focus on units.

Concentrating on Units

Although we consider arsenic in rice, worldwide the risks from ingesting arsenic come primarily through drinking water. In the US, arsenic concentrations in drinking water are limited by the EPA to a *maximum contaminant level* (MCL) of 0.01 mg/L or 0.01 ppm ("parts per million"). One-hundredth of a part per million means one part in 100 million or 10 parts per billion, 10 ppb.

> *Do you understand these units?*
>
> *Think of your response and jot it down.*

I didn't. The unit mg/L seems to say we measure the amount of arsenic in milligrams and divide by the number of liters of the sample of (not necessarily pure) water. In fact, that's what it means, and it is simple enough. But the ppm and ppb tripped me up. I mean, a part per billion *seems* understandable enough. If there is a giant bin of a billion tennis balls and all are yellow except for one red ball, then there is one red per billion. If the bin has twice as many balls and two are red, then that's two parts per two billion (recall that "per" means "divided by"), or again one part per billion, 1 ppb. Ten red balls would mean 10 ppb. Note that ppb is a ratio of numbers, so it has no units. But what does this mean when we're talking about water? That is, we have a solution mostly of water but with a small amount of the "contaminant." The "parts" in question are the contaminant – in our main question, the molecules of arsenic.

Returning to our warm-up example, let us ask, what if the contaminants, the red tennis balls, have different volumes and different masses than the others? Then ppm could be sensibly interpreted in at least three different ways – number per number, volume per volume, or mass per mass:

1 number of red tennis balls out of a million total tennis balls
2 volume of red tennis balls divided by volume of total tennis balls, times a million
3 mass of red tennis balls divided by mass of total tennis balls, times a million

The first is the most natural, since we think of "parts" as referring to a *number* of objects. Now number 1 agrees with our discussion above. Note that this can also be phrased as the number of red tennis balls divided by the total number of balls times a million: one red ball in a million would give a ratio of 1/1,000,000 but means 1 ppm, so we have to multiply the ratio by a million. That's why items 2 and 3 have "times a million" at the end. Now the reason for the ppm unit rather than the straight ratio is to avoid tiny numbers when the concentrations are weak, which is often the case.

Moving on to number 2, what does it mean? First, let's suppose that the volume of a red ball is the same as the volume of a yellow ball. Then the ratio of volumes would be the same as the ratio of numbers. So if there were one red ball in a million, then the straight ratio would be 1/1,000,000. But if we had the same total volume of balls and the volume of our single red ball were *twice* the volume of a yellow ball, the ratio of volumes would be larger than it was for numbers: so 1 and 2 would give different answers. Moving on to number 3, we see this is similar to 2, except we're using mass instead of volume. So even if we had just one red ball of the same size and volume as the yellow ones, if it were more massive, then the ratio of masses would be different. So these three different notions converge only when the red balls have the same mass and size as the yellow ones.

We want to use one of these notions to describe the amount of a contaminant in water, or perhaps more generally, to describe the amount of a *solute* (such as arsenic) dissolved in a *solvent* (such as water) to create a *solution*. It turns out that for measuring the level of a contaminant, since it is easiest to measure the total *mass* of the solute (say, arsenic), option 3 is used. Try to figure out the concentration expressed in ppm and ppb for the following example:

A 5 kg canister of paint containing 1 mg *of mercury.*

Think of your response and jot it down.

It helps to know that one kg equals a million mg, since $kg = kg \times \frac{1000g}{kg} \times \frac{1000mg}{g}$.
So we have 1 mg of mercury in 5 million mg of paint, for a concentration of
$\frac{1}{5}$ ppm $= 0.2$ ppm $= 200$ ppb.

If you could complete the exercise, then you should now have a good idea of
what ppm measures. Now the unit mg/L is self-explanatory (mass of solute in mg
divided by volume of solution in liters), but how can mg/L and ppm ever be the
same? How can we possibly reconcile the two when they're not even the same *kind*
of unit!? For example, ppm represents a ratio of two masses, so the units "cancel."
It's just a pure number. But mg/L is a mass (milligrams) divided by a volume (liters).
Here's the crucial fact that brings it home:

a liter of water has a mass of one kilogram.

So *if* the solution weighs the same as water, then we can equate 1 mg/L with
1 mg/kg, in other words, 1 milligram per million milligrams: 1 ppm. Summarizing,
under these conditions,

$$1 \text{ ppm} = 1 \text{ mg/L} = 1 \text{ mg/kg}.$$

We can easily visualize what a part per million – one milligram per kilogram –
might look like. A kg of water occupies a liter, which is the volume of a cube
of length 10 cm or 100 mm. So it has volume $(100 \text{ mm})^3 = 1,000,000 \text{ mm}^3$. This
means there are a million cubic millimeters in a liter. (See Figure 4.1.)

Loosely speaking, a cubic millimeter is about the size of a "crumb." So if your
crumb has the same density as water, then one crumb in a liter of drinking water
would "contaminate" it with a concentration of 1 ppm. This crumb is comparable to
the amount of food that you feed to your fish in your 100-liter tank, meaning the
fish's food "contaminates" their water at a concentration of 1/100 ppm or 10 ppb.

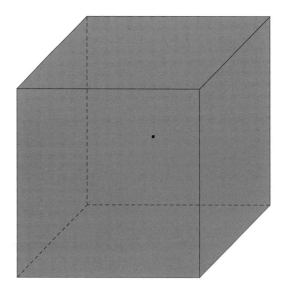

Figure 4.1 The large cube has volume a million times that of the small cube in the center.

In our reasoning, we said "*if* the solution weighs the same as water," so let's be clear here: equating one liter of potentially contaminated water with 1 kg rests on the following:

> *Assumption*: the levels of contaminant in the solution are low enough that the overall mass is about the same as the mass of pure, uncontaminated water.

Since we are measuring the contaminant levels in parts per million or billion, this would appear reasonable. For instance, if you have a *million* people and all have hair between 4 and 6 inches in length, except for a dozen or so outliers with hair down to the ground, they will not affect the average noticeably. The analysis is similar if we're talking about the mass of molecules and a few of them are heavy arsenic molecules compared to lighter water molecules. Still, when there is an assumption, there is always a chance it is off, so we must be mindful that we are assuming that our solution is sufficiently *dilute*, i.e., a liter of it has about the same mass as a liter of water, 1 kg. Then $1\,mg/L$ is the same as $1\,mg/kg$ or 1 ppm.

To see how this can go wrong, suppose the "contaminant" is sugar dissolved into water. At room temperature you can dissolve about 2 kg of sugar in 1 kg of water. So you can have a liter with $2\,kg = 2{,}000{,}000\,mg$ of the contaminant. This would produce a concentration level of $2{,}000{,}000\,mg/L$, but if we convert this to ppm using the rule that $1\,mg/L = 1$ ppm, we get a level of 2,000,000 ppm, or *two million parts per million*! Huh? That makes no kind of sense! The problem is that our assumption is not valid in this case.

The long detour we took to understand the units was necessary. Put frankly, if we didn't understand the units we were using, then at some level of scrutiny, we'd have to admit that we didn't know what we were talking about!

Arsenic in Rice

So now we have an idea of what inorganic arsenic (As) is and how its levels are measured, and we are sure that it is dangerous when appearing in high levels (>10 ppb) in drinking water. That's all well and good, but the question we asked was about arsenic in *rice*.

Remark 4.1 After all this research on dangerous arsenic levels in drinking water, I was curious about the water supply here in my town of Evanston, Illinois. A search for "Evanston water report" quickly led to a document called the "Consumer Confidence Report," which contained data on levels of sodium, chlorine, coliform bacteria, and other substances in the water – however, no mention of arsenic. There was a phone number listed for consumers who had questions (the city's 311 information line, in fact), so I called and left a message. Later that day, I got a phone call back explaining that there was not enough arsenic to be detected, meaning that the levels were less than 1 ppb. I felt lucky that not only was my drinking water so safe but my city has allocated sufficient resources for its water utilities and to provide an information hotline. Many are not so lucky.   ▲

Does it follow from the analysis of arsenic in water that it is equally dangerous when it appears in rice? It would seem so, but scientists cannot make such assumptions and instead must gather data. Our bodies process rice and water differently, and so the two could have different effects. For a silly example, note that kids who eat ice cream are happy. So, do parents who buy ice cream at the store have happy children? Perhaps – but we'd have to ensure that the ice cream gets from the parents to where it needs to go: the children's mouths. More seriously, suppose you take vitamin supplements. They will not do any good if they are not absorbed by your body, and most vitamins are better absorbed with fat and protein. So if you don't take them with a meal, they might not have the desired effect.[5] Likewise, certain compounds may cause harm only in combination or in particular environments. Without research, we cannot know for sure if arsenic in rice has the same effect as arsenic in water.

So let us ask,

Is there specific knowledge of the link between arsenic in rice and cancer?

Clearly we cannot address a question like this by performing controlled experiments where we feed people high levels of a suspected carcinogen. Yet there are populations for which rice figures prominently in the diet, so we have valuable epidemiological data.

At the time of writing, the FDA is currently investigating the long-term health effects of arsenic in rice. While it found that the short-term risks were not a cause for concern, no conclusions have been drawn about whether consuming rice regularly for long periods of time can eventually cause harm.[6]

Possibly because of this doubt, at the time of writing, the FDA sets no limits on the level of arsenic in food. (As we remarked in a prior footnote, the lack of action could be due to political pressure from one side, or even *lack* of political pressure from the other.) That said, the Food and Agriculture Organization (FAO) of the United Nations in 2014 approved the guideline that arsenic levels in rice should not exceed 0.2 mg/kg (note that since rice is not liquid, we use a mass measurement for both the quantity of arsenic *and* the quantity of rice).[7] This limit is 20 times the limiting level in water set by the EPA ($0.2/0.01 = 20$), so there'll be as much arsenic in 100 g of rice at this level as in 2,000 g of water, or two liters. A "serving size" is about 75 g of uncooked rice, so you might ingest more arsenic through eating rice every day at the FAO contamination limit of 0.2 mg/kg than through drinking water at 0.01 mg/L.

---

[5] And even then, we cannot just assume that more is better; and perhaps we don't really know which vitamins actually do what. Maybe they only work in combination with other factors that are not present in the supplement. And so on. There are many questions.

[6] The source for these data is unfortunately now a dead link: www.fda.gov/ForConsumers/ ConsumerUpdates/ucm352569.htm. To view this and other now-defunct websites, first go to the Internet Archive's Wayback Machine at https://web.archive.org, then enter the URL there.

[7] www.fao.org/news/story/en/item/238558/icode/.

Now the rice on American supermarket shelves may not be at the FAO limit, so we should look at the data. After all, many municipalities (such as Evanston) have levels of arsenic in their water supply much lower than the MCL, so when it comes to water we are not necessarily in the worst-case scenario. The same may be true of rice, so let's press on and study what the levels of arsenic in food actually are.

There are lots of kinds of rice: short grain, medium grain, long grain, sushi rice, and the aromatic rices jasmine and basmati. And there are brown or white versions of all of these. Furthermore, these varieties can be grown in different regions, and different companies have different manufacturing processes and different suppliers even within the same region. To get a meaningful assessment – you don't want a single bad sample skewing your conclusions – you must test a number of samples from each variety. Once again, the problem becomes a can of worms! Here we won't try to report all the efforts of the Food and Drug Administration (FDA) or *Consumer Reports* but will simply present some of the data.[8]

Rice	As per serving (µg)	mg/kg
basmati	3.5	0.078
brown	7.2	0.16
instant	2.6	0.057
jasmine	3.9	0.086
short	3.5	0.077
medium	3.6	0.079
long	4.6	0.10

The FDA reported food levels by serving size. A serving size of dry rice was defined as 45 g (the FDA and I disagree on serving size, but then again I do love rice). Note that a µg ("microgram") is a millionth of a gram, or a thousandth of a milligram.

> *How did we generate the* mg/kg *column from the FDA data?*
>
> *Think of your response and jot it down.*

To answer this question, we do a units conversion:

$$1\frac{\mu g}{\text{serving}} = 1\frac{\mu g}{\text{serving}} \times \frac{1 \text{ mg}}{1{,}000 \text{ }\mu g} \times \frac{1 \text{ serving}}{45 \text{ g}} \times \frac{1{,}000 \text{ g}}{\text{kg}} = 0.0222 \text{ mg/kg},$$

so if 1 µg/serving = 0.0222 mg/kg, then for basmati rice, 3.5 µg/serving = 3.5 × 0.0222 mg/kg = 0.078 mg/kg, and similarly for the other varieties.

> *In the right column, why did I write 0.078 instead of 0.08 for basmati rice, and 0.10 instead of 0.1 for long grain?*
>
> *Think of your response and jot it down.*

[8] www.fda.gov/Food/FoodborneIllnessContaminants/Metals/ucm319870.htm.

The reason is that the data were given to two significant digits (middle column), so the converted data should be given to the same degree of precision.

The above data from the FDA do not differentiate between different regions and only tested one type of brown rice, as compared to the *Consumer Reports* data. We will use this for our purposes, but if we consider ourselves to be *California* rice eaters, for example, we would need the more detailed data set.

The table shows that brown rice has by far the most arsenic. The reason, according to the reports, is that more arsenic is stored in the outer layers of the grain, which remain in brown rice but are removed in the making of white rice.

If we are not eating brown or instant rice, the remaining choices do not differ greatly. The average of these five numbers is 3.62 µg per serving, or 3.6 µg per serving, as we are only retaining two significant figures. So let us consider ourselves to be an "average" rice eater, eating white rice with this level of arsenic, and let us now decide how many servings we plan to eat each day. This will vary greatly from person to person, but let's pick a number: three servings per day? Personally, I don't eat rice every day, maybe every third day, but when I do, I like to eat a lot of it, maybe even nine servings. At a restaurant, I get very anxious that I won't get enough rice. I even asked my German friend to create a single word for my phobia: *Restaurantsreisunterfütterungsangst*! At three servings per day, we're talking about 3×3.6 µg $= 11$ µg of arsenic.

(Of course, when we make rice, we use water, and that may also contain arsenic. Since the city of Evanston records no measurable amount of arsenic in its water supply, I shall not include this consideration in my analysis. Reader: you may. Check local listings.)

Though the FDA does not set guidelines for arsenic in rice, the EPA standard for arsenic in drinking water was a limit of 10 ppb $= 0.01$ mg/kg $= 10$ µg/kg $= 10$ µg/L. We just found that at the assumed rate of rice eating, we would consume 11 µg per day of arsenic, so this is the same amount as we would get drinking a little more than a liter of water at the highest acceptable level. Now since it is reasonable to assume that some people will drink an extra liter of water a day, there does not seem to be serious cause for concern.[9] Brown rice had twice the arsenic level as white (7.2 µg vs. 3.6 µg), but two liters of water a day might not be so unreasonable either. (Brown rice has other benefits, perhaps, such as being higher in fiber, but on the arsenic tip, it's a clear loser.) Finally, keep in mind that the link between arsenic in food and cancer is not considered firm.

Decision

Since eating rice aggressively at three servings a day leads to an extra amount of arsenic within the normal range of what one might consume drinking water, I

---

[9] Again, this is only based on the total amount of arsenic ingested; we have not accounted for the different ways our bodies might absorb arsenic in rice versus water, because we simply do not know.

will continue to eat rice without concern. While I will continue to worry that I'm not getting enough of it when at restaurants, I won't worry about the cancer risk. Others may be more cautious. You may draw your own conclusions, but whatever they are, they will be informed choices based on quantitative reasoning!

Further Topics

We have discussed some basic probabilities, such as the likelihood of developing cancer, and mentioned statistics in the form of averages and correlation, though we didn't get into details. After all, we're not applying for a job at the NIH – well, not just yet. We will discuss some of these issues in more depth in Chapters 6 and 10. For now, though, it is worthwhile to note some further topics that one can explore.

Confounders

Suppose that in the sampled water wells in Taiwan there is, in addition to arsenic, an undiscovered vitamin that helps *prevent* lung and bladder cancer. Then we would conclude that the necklaces had an effect: the seriousness of arsenic is even *more* alarming, since the rise without that vitamin would have been even more significant. Or suppose, as may be more likely, that the water tables contained something *besides* arsenic that has a similar ill effect on health. Then the dangers of arsenic would be *less* than we had thought.

Such a factor – an extraneous statistical variable that correlates (either positively or negatively) with the ones measured – can "confound" the analysis and so is called a *confounder*. Controlling for these effects requires more advanced statistical theory.

Attributable Risk, NNT, NNH

Suppose three in ten people who went out got the flu, but four in ten people who went out *with necklaces* got the flu. If we were able to eliminate all other confounding factors from these data, we would conclude that the necklaces had an effect: the *attributable risk* due to wearing a necklace is that extra one instance of flu, or 1 in 10. It is the difference in probabilities with or without exposure to the "intervention" (the necklace). Put differently, upon sending ten people out with necklaces, we chalk *one* of those flus up to the necklace-wearing. The *number needed to harm* (NNH) here is 10, the number (of necklace wearings) needed to cause one extra case (of the flu). NNH is computed by taking the reciprocal of the attributable risk.

If we're talking about helpful drugs that decrease the risk, then we talk about the number needed to treat (NNT), and it is the reciprocal of the attributable decrease in risk.

Summary

Let's review. We found that painstaking scientific research identified a correlation between high levels of inorganic arsenic in drinking water and cancer. Levels of contamination are reflected by concentrations of solutions, and we spent time ensuring that we understood the units used in discussing concentration.

While the FDA sets no safety limit on arsenic in food, nor has it conclusively linked arsenic in food to cancer, we found what we thought to be a good measure to regulate consumption of arsenic through rice: the amount of arsenic that might be consumed in a day by drinking water at the MCL of the EPA. We then used data for arsenic in various rices and decided on the amount and type of rice we were intending to consume. We found that eating white or brown rice regularly could be equated with drinking about one or two liters of water at the MCL, respectively.

With this knowledge in hand, each reader is informed enough to make a choice. The author loves rice and found the levels low enough not to warrant concern (warning: potential emotional bias here!) but leaves the reader to draw his or her own conclusion and make his or her own decisions.

Beyond rice (as important as it may be to the author), the point of the chapter is to learn to apply the same methods to decisions about any other potential risks that we encounter in our daily lives. Further examples are explored in the exercises and projects below.

Exercises

1 Chlorine levels in pools should be kept around 2 ppm. Your pool is 18 ft by 36 ft, with a depth ranging continuously from 3 ft to 7 ft. What is the mass of chlorine in your pool?

2 Your five-year-old cousin takes swimming lessons and gulps a mouthful of pool water each time. What amount of drinking water at the maximum contaminant level (MCL) of the EPA would contain the same amount of chlorine? The MCL for chlorine is 4 mg/L.[10]

3 The MCL for mercury (Hg) is 2 ppb. If someone put a drop of mercury in a pool just filled with clean drinking water (no chlorine) and offered you $1,000 if you took a sip from it, would you do so? Explain.

4 From 2013 to at least 2016, the city of Flint, Michigan, experienced a crisis of lead contamination in its water supply. Research the values of lead observed in the water at various times and compare with official guidelines for safe water. Compare with values measured in your hometown (if different from Flint) during the same periods. Comment on what you can conclude from your data. (This exercise involves research.)

5 In chemistry, concentrations are often measured in terms of moles per liter, mol/L, or molars (M). (One mole of a substance is equal to 6.022×10^{23}

---

[10] http://water.epa.gov/drink/contaminants/index.cfm#Disinfectants.

particles of the substance. You will not need to use this fact.) The mass in grams of one mole of a substance is equal to the mass in atomic mass units (u) of one particle. The atomic mass of arsenic is 74.92 u. Convert the EPA's MCL for arsenic in water of 10 ppb to molar units.

6 The city of Beijing experienced Red Alerts for high levels of smog at various points in 2015. Find out the levels for one such alert and compare with what is considered a dangerous level of smog by your government. Calculate how much fine particle pollution, or $PM_{2.5}$ (for "particulate matter" less than 2.5 micrometers in diameter) you would breathe in at those levels in a single day. (This exercise involves research.)

7 Which is more dangerous, driving or flying? Specifically, taking only the risk of death into account, is it safer to be driving in a car or flying in an airplane for one hour?

   Hint: Here are some helpful numbers: in 2013, there were 27,241 traffic deaths (excluding pedestrians and cyclists) in the US.[11] Americans drove 2.972 trillion miles that year.[12] Since the accidents were divided roughly evenly between rural and urban roads, we can estimate the average car speed to be around 45 mph. Also in 2013, there were 62 airplane fatalities in the US.[13] Americans flew 589.7 billion miles domestically that year.[14] The average speed of a commercial jet is 550 mph. Note that we are comparing equal *times* of travel, not distances.

8 Suppose a large study reveals that 6.1% of people who go out without a sweater catch a cold, compared to 5.7% of those who wear sweaters. What is the "number needed to treat," i.e., how many times do you need to wear your sweater to say that it kept you from catching one cold?

9 You and 89 classmates go on a school camping trip for the weekend. After the trip, 15 out of the 60 of the people who went swimming in the lake over the course of the weekend became sick with the flu. You did not go swimming, nor did you get sick. However, you know that 6 out of the 30 people who didn't go swimming also became sick with the flu. Calculate the attributable risk of catching the flu from swimming in the lake and the number needed to harm (i.e., the expected number of swims that would result in one additional case of the flu *due to swimming in the lake*).

10 Think of a topic for a project related to this question.

---

[11] www-nrd.nhtsa.dot.gov/Pubs/812101.pdf.
[12] www.fhwa.dot.gov/policyinformation/travel_monitoring/13dectvt/13dectvt.pdf.
[13] www.baaa-acro.com/advanced-search-result/?year_post=2013.
[14] www.rita.dot.gov/bts/sites/rita.dot.gov.bts/files/publications/national_transportation_statistics/html/table_01_40.html.

Projects

A Lead can lead to impaired mental development in children. You just found a half a paint chip on the ground outside your old shed, which was painted with lead-based paint back in the day. Should you worry? Be sure to compare your concerns with those of the residents of Flint, Michigan (see Exercise 4 above), or of the West Calumet Housing Complex in East Chicago, Indiana.

B Radon gas is the leading cause of lung cancer among nonsmokers. How worried should you be about radon in your home?

C Would you let your child play high school football? Create a quantitative analysis of the question to formulate your response. (Don't just write "No, I hate football.")

5

What Is the Economic Impact of the Undocumented?

Appendix Skills: Estimation, Probability

OVERVIEW

This chapter addresses the sensitive question of unauthorized immigrants in the following five steps:

1 *First, we acknowledge the emotional impact of the question. We know that economic considerations cannot be the whole story, but we engage in the exercise of trying to measure them.*
2 *Next we opt for a cost–benefit analysis from the perspective of federal and local governments in the US – then try to think of what kinds of items to tally.*
3 *We do research, but face problems: sources can be biased, and even nonpartisan sources cannot get good data on the undocumented population. We make assumptions to bridge the gap between the data we have and the data we need. Still, unknowns remain as large as knowns.*
4 *Our tally of known and estimated costs and benefits comes close to even. We conclude that we cannot justifiably place economic considerations at the forefront in considering immigration and the undocumented.*
5 *We end by affirming that whatever policies we support must be consistent with basic decency.*

Reactions and Reflections

What are your first impressions when you read a question like the one posed in the title for this chapter?

Think of your response and jot it down.

The question is one of those political hot-button issues.[1] Immigration, legal or not, is always a fraught question for residents of a country, mixing as it does feelings about nationality, race, socioeconomic status, the economy, border and homeland security, xenophobia, and many other sensitive topics. One thing to know immediately when trying to wrap your head around a question like this is that

---

[1] This chapter was written in 2015. In the subsequent years, the question of immigration became even more of a political flashpoint.

you and almost everyone else will respond emotionally. But some will react with unbridled emotion, that is to say, outside of the scope of reason and analysis – and those people are unlikely to be swayed by any response that runs counter to their pre-formed opinions.

Consider: if the title of this chapter were "Do Illegal Immigrants Cost Me Money?" that phrasing could likely be interpreted as a provocation. Many note that *actions* may be illegal, but not *people. Undocumented immigrants* is currently the preferred term, but there is ongoing debate.[2] If we had used the coarser phrasing, a reader without documentation, or with an undocumented family member or friend, might reflexively distrust the author. Another issue with the alternate question is that it assumes that the asker "me" is not undocumented and that the undocumented are the "others." Even a more neutral phrasing like the one in this chapter will engender reactions, e.g., by suggesting that we should prioritize economics when considering immigration. On the flip side, to a nationalist, the "correct" phrasing used in the question may be seen as pro-immigrant and elicit a knee-jerk response. These are sensitive issues, but that is no reason to shy away from the subject.

You should also try to be aware of whether you yourself will try to justify a pre-formed opinion.

> *What potential biases do you bring to a study of immigration?*
>
> *Think of your response and jot it down.*

Addressing potential biases is an important part of quantitative reasoning. A bias could be anything that causes an unfair result or "tips the scale." Parents know how to bias a question to a child: "Do you *really* want to go to the Italian restaurant [frowning], or do you want to come with all of us to the Chinese restaurant [smiling]?" This example illustrates intentional bias, but biases can often be unintentional or unconscious.[3]

In making a quantitative argument, you should acknowledge the role of emotionality. Quantitative reasoning may give you a firm basis for your argument, but if you barge into your friend's living room to present new evidence to her that chocolate milk is healthy, your rigor will fail you if you have insulted her by not removing your shoes.

That said, how does one start forming an unbiased opinion about a question that can be addressed along so many different lines? It's nearly impossible. To see why, let's ask a simpler, sillier question: is it better to take a sweater with you when you go out, or not?

If it gets cold, you'll want the sweater. If it doesn't, you'll have the nuisance of schlepping it along with you. Different individuals will value staying warm and

---

[2] See, e.g., www.huffingtonpost.com/2013/04/02/ap-drops-term-illegal-immigrant_n_3001432.html. As I write this, that link is aging. In many circles this is a settled matter, but not all. How offensive a term is changes in time, as well: consider the "C" of NAACP or the "Q" in LGBTQ.

[3] This site has some fun, but serious, tests of implicit bias: https://implicit.harvard.edu/implicit/.

not having to carry things differently. Some don't mind tying their sweater around their waist if it means insurance against feeling cold, while others wouldn't mind a potential chill if it'll let them walk around town unfettered. Each person, then, has their own risk–reward analysis.

There is some risk of getting chilly measured against the reward of walking emptyhanded. Even if the valuation of these measures is emotional – "I hate being cold" or "I love walking in my T-shirt" – we can approach the decision quantitatively and *objectively*, but we must openly admit that the valuations themselves may have been *subjective*.

Quick story: when I was a high school senior choosing which college to attend, I rated each school on several points – reputation, location, social life, etc. – then I weighted each score based on which trait was more important to me (that is,

---

Tactics in Argumentation

Don't fall victim to these rhetorical devices, convincing as they may seem.

- Straw man (mischaracterizing or distorting an opposing argument). *I'll never go vegan; cheese does <u>not</u> have a soul!*
- Reduction to the absurd (logical extreme). *If we all went to art school, the economy would fail.*
- Cherry-picking ("proof" by example/ anecdote). *My mom smoked a pack a day and lived to 93.*
- Appeal to authority. *How could New York be so great if Warren Buffet lives in Omaha?*
- *Ad hominem* attack. *You stand there with <u>those</u> <u>shoes</u> and try to tell me about global warming?*
- False dichotomy (asserting the two options presented are the only ones). *Either you fund this fighter jet or you cede all control to the enemy.*

You can build up immunity, and credibility, by quickly recognizing these and other devices (leading question, emotional appeal, slippery slope, vague language), then demanding that the argument be made on its merits.

I multiplied each score by an "importance" factor), and I added up the scores to compare schools. One school came out on top, but I promptly rejected it, threw away the chart, and went with my emotional choice!

In that scenario, I allowed for emotionality to drive the response, in the end ranking the feel-good factor above all others. But there are many situations where including an individual's subjective biases is not acceptable. If you are the chief financial officer of a company, you will want to restrict your attention to how your choice will affect the "bottom line," not what "feels best." If you are a policy analyst, you will make recommendations for policies that will affect many people: each will have their own quirks and opinions, so you must avoid all emotional biases and decide the question purely on its merits. You also have to insulate yourself against the tactics of persuasive arguing (see sidebar). But how is this done?

It's important to *decide how you will decide* a question before making the actual decision, otherwise you can rig the outcome. Maybe you start your sweater analysis by saying that remaining unfettered is twice as important as staying warm, then you look up the weather forecasts and probabilities of cold spells and make your analysis and the numbers tell you not to bring the sweater. But you don't like the sound of that, meaning you actually think you'll wind up unpleasantly cold. If this happens you *could* go back, rig the answer by changing the weighting to value warmth and unfetteredness equally, and engineer the answer that you want. Now perhaps the first answer didn't feel right because in fact you had misvalued how much you dislike the cold – and for a trivial analysis like sweater-carrying, it's no big deal –

> **Emotions versus Reason: An Exercise**
>
> You show up at your first day of chemistry class. Your instructor walks in and starts to speak. You immediately perceive their gender, age, race, ethnicity, style, accent, manner of speaking. You also notice the physical space, the other students, demographical patterns, attitudes, and seating choices. You may have an instinctual, emotional reaction to each of these observations. That is perfectly normal and part of human nature. It would be a shame, however, if these reactions had a negative impact on your performance in the class, which should depend on your reasoning skills. You must train yourself, if necessary, not to let the fact that "my math professor is a total dork" impact your commitment to your education. Maybe the dork at the front of the class is capable of fascinating observations about the subject. You need not have a personal relationship with your instructor to have a productive professional one.
>
> Should reason or emotion drive these choices?
>
> - Which restaurant?
> - Advocating higher or lower taxes?
> - Blues or Jazz?
> - Selecting an electricity carrier.
>
> See, it seems clear once we put it like that – but questions like the one in this chapter elicit emotional *and* rational responses, and we need at least to distinguish between them in our analyses.

but it doesn't look good to make substantive decisions in such a haphazard way. Surely, an independent reviewer of your process would take you to task for unduly influencing the results.

Sweaters are not matters of national importance (well, in the wool-producing country of New Zealand, they are quite important, and people's opinions about them may be affected by their national pride or their desire for economic stability), but immigration *always* is. If you want an honest, independent analysis of your question, decide *in advance* how you will approach the quantitative analysis.

So let's get down to the business of addressing the question of the economic impact of the undocumented population.

Spoiler alert! We won't get a clear answer. The question is just too thorny. Scores of scholars working for years at think tanks have not settled this question, but there is a lot to be learned from trying.

Regarding "impact," we should ask, on whom? On the US government? An average citizen? The state of Illinois? We could take a number of approaches, but let's make a concrete choice.

We will measure the costs and benefits based on tax revenue and government expenses.

Again, the choices and assumptions we make are important: if we study the impact on a major Californian fruit grower, the calculus of the answer could change significantly, since undocumented immigrants are employed as fruit pickers in numbers disproportionate to their percentage of the population.[4]

What does that last sentence mean?

Think of your response and jot it down.

It means that the percentage of fruit pickers who are undocumented immigrants is higher than the percentage of people living in the US who are undocumented immigrants.

Okay, so we will try to dispel our gut reactions and conduct a sober analysis of the governmental expenses and revenues associated to the undocumented population in the US.

Identifying Revenues and Expenses

There are many ways to begin. ("Are we *ever* going to begin?" you may be thinking – but all this preamble is a big part of it. In other words, we've already begun!) Here, since we're discussing a question of cost, we should add up the ways in which undocumented immigrants contribute to the revenue of the government (benefit) and subtract the ways in which they receive services that come at an expense (cost).

If revenue minus expenses is positive, what does that mean? What if it is negative?

Think of your response and jot it down.

---

[4] There is an article on the website of the German broadcaster Deutsche Welle, www.dw.de/imroving-working-conditions-for-illegal-fruit-pickers-in-california/av-15601612, citing the United Farm Workers Union (UFW) as estimating that 60% of the roughly 30,000 farm workers in California are undocumented immigrants. I could not find such a statement at the UFW website. However, the Pew Research Center, a nonpartisan source, reported that about 34% of farm workers in California are undocumented immigrants (www.pewhispanic.org/2015/03/26/appendix-a-state-maps-and-tables/) – a smaller fraction, but still much larger than the fraction of the overall population represented by the undocumented.

Revenues

Now before we calculate the total revenue due to undocumented immigrants, we should list the different revenue streams or ways in which immigrants contribute to the economy as measured in tax revenue – for while there will certainly be other contributions, we have decided here to focus on receipts.

> *How many ways can you think of?*
>
> *Think of your response and jot it down.*

There is always a danger that we may miss some, but here are some ways:

1 their own individual tax payments
2 the taxes of their employers
3 the sales taxes from their purchases
4 property taxes
5 other secondary effects

Let's discuss.

Number 1 is straightforward to understand. Undocumented immigrants pay taxes. We will have to find sources of data for this and all the questions. Knowing what data you require is quite a stretch from *having* said data!

Number 2 is easy to understand but will be even more difficult to measure. Employers who hire undocumented immigrants pay taxes on the revenue that their employees help generate, but how much of those taxes can we say are *attributable* to undocumented workers? This will be a toughie.

As for 3 and 4, undocumented immigrants spend money and, with each purchase, pay some sales tax. How might we measure this revenue? If we are lucky, we will find someone who has already calculated it, but otherwise, we might have to figure out how to access those data indirectly. To look more closely, we will spend some time asking,

> *How might we measure how much of the government's sales and property tax revenue comes from the undocumented?*
>
> *Think of your response and jot it down.*

Let's think of the question this way. Suppose a state receives a dollar of sales tax. What is the chance it came from an undocumented immigrant? If we knew the answer, then that fraction of the total sales tax revenue would give us the datum we seek. A first guess for this fraction might be the total number of undocumented immigrants divided by the total number of people, since this is the probability that a randomly chosen person is undocumented (see Appendix 7).

> *What assumptions does this guess implicitly make?*
>
> *Think of your response and jot it down.*

It assumes that undocumented immigrants make purchases at the same dollar rate as documented residents and citizens. Consider the following:

$$\frac{(\text{sales tax per undocumented resident}) \times (\text{\# undocumented residents})}{(\text{sales tax per resident}) \times (\text{\# residents})} = \frac{\text{sales tax from undocumented residents}}{\text{sales tax from residents}}.$$

So *if* the sales tax per undocumented resident equals the sales tax per resident, then the very left terms in the numerator and denominator on the left cancel out, and the ratio of taxes paid by the undocumented to taxes paid by general residents is equal to the fraction of residents who are undocumented.

> *Why might the assumption that undocumented immigrants generate taxes at the same rate as documented residents be valid or invalid?*
>
> *Think of your response and jot it down.*

There are several ways in which this assumption can fail. For example, undocumented immigrants generally have a lower income, meaning they have fewer dollars to spend – so counting the *number* of undocumented residents and converting using the sales taxes paid per person will not be as fruitful as trying to measure their total *income* somehow and finding the US sales tax collected per dollar of total income.

One way of getting a more accurate estimation is by measuring the income from sectors of the population that are demographically similar to that of the undocumented – a more sophisticated analysis than we have proposed so far. Our naïve assumption may lead to an overestimate of tax revenue from the undocumented, since their per-person sales tax generation may be lower than average. On the other hand, they may have less access to lines of credit and savings, and live closer to the margins, meaning they may be more likely to spend the dollars they *do* have, and the closer demographic analysis should account for this discrepancy as well. For now, we will simply record this assumption and move on.[5]

Remark 5.1 In fact, in preparing this chapter, we ran the numbers from this naïve assumption and found them to be nearly *twice as high* than the number created by think tanks – precisely because of the discrepancy discussed. Had we wanted to "cook the books," we could have used this overestimate. The lesson is that not all assumptions are as gentle as they appear, so quantification of error is important. As stated previously, in researching this chapter, we focused on "nonpartisan" sources only, for indeed other organizations have reports on the subject filled with obvious overestimations of whatever data support their mission. ▲

What other effects referred to in item 5 might there be? Laborers in the workforce make and do things. This means more goods and services available to the public, who can use these to produce more wealth and revenue. Earners spend, and spending creates demand for new products. All of this means more money and jobs in the economy. Workers also make money for their employers, who spend and pay taxes

---

[5] Note we are moving on not because we have resolved the question of the merits of this assumption but because we want to develop our model. In fact, we will later find data that measure receipts rather than trying to estimate them crudely, so this estimation will be abandoned anyway.

due to the new revenue, as in item 2. We could label all of these as downstream effects of undocumented laborers, arguing that the effect of additional workers is nonlinear, similar to interest on interest (see Chapter 3). It may be the case that undocumented immigrants are so integrated into the economy that our model of an economy in which you can isolate their contributions is flawed from the get-go. How can we tell? We will return to this point in short order.

Having detailed the various ways in which undocumented immigrants contribute to the US treasury, we must now turn to ways in which they are an expense to US taxpayers (which includes the undocumented themselves).

Expenses

Estimating the costs associated to undocumented residents is tricky as well. As we have already discussed, due to demographic differences between the undocumented immigrant population and the general population, we cannot assume that the average undocumented immigrant receives the same benefits (hence costs the Treasury the same amount) as the average US citizen. Besides, many entitlements, such as Social Security, require proper authorization, so they may not be disbursed to the undocumented. Therefore, simply multiplying the total amount of government money spent in a year by the fraction of undocumented immigrants in the US population will give an answer prone to significant error, just as it did for revenue estimation (see Remark 5.1).

So how should we begin enumerating costs? We could identify the government expenses from which undocumented immigrants are most likely to benefit and use demographic analyses or direct data to gauge the economic costs they impose.

> *Can you think of ways in which undocumented workers might benefit from taxpayer money?*
>
> *Think of your response and jot it down.*

The US Congressional Budget Office, in its 2007 report on the *Impact of Unauthorized Immigrants on the Budgets of State and Local Governments*,[6] cites three areas of government spending from which undocumented immigrants are most likely to benefit:

- education
- health care
- law enforcement

Of course, there are more areas in which undocumented immigrants may benefit from taxpayer money, such as police and fire protection, paved roads, and other municipal services. So we should add another category:

- other

---

[6] Congressional Budget Office: www.cbo.gov/sites/default/files/12-6-immigration.pdf.

When we gather data on these costs, this last category will introduce a large error, similar to the secondary effects on revenue.

Data: The Gathering

It is now time to start gathering the data we'll need. We'll see that it gets messy.

Revenue

We have a rough idea of the *kinds* of contributions undocumented immigrants make to the economy, but we won't have any idea of the *amount* of these contributions without some actual research.

The gathering of data can be messy: surveys only cover portions of the population, contain inaccuracies and misreported information, and may not address the question you want anyway. Government data are typically better but are not necessarily designed for the purposes you require. A lot of the skill of being a good economist, clinical psychologist, historian, or what-have-you is designing ways to create or procure clean data. We have not done this systematically, which means *this is not a scholarly report on the economic impact of undocumented residents*! Instead, it is a demonstration of quantitative reasoning, in this case, as applied to a question of immigration economics.

Though we are only getting our feet wet as scholars, we made an effort to search the Internet and other literature for clean sources. In an all-roads-lead-to-Rome situation, we found that news articles and other sites we stumbled upon made repeated references to just a few particular nonpartisan groups or government agencies. After many misleads and click-throughs, these are three items we found most useful:

- the Institute for Taxation and Economic Policy (ITEP) report[7]
- the Congressional Budget Office (CBO) report[8]
- the US Census[9]

The number of references we encountered was a clear sign that the figures of these institutions were respected.

The ITEP report was a comprehensive accounting of state and local tax receipts from undocumented residents. What about federal taxes? One news article on tax payments from undocumented workers referred to a staff member at a think tank, whom we contacted for leads on federal tax payments of the undocumented: he replied![10] I learned that this was murky water even for experts – cold comfort. But then I found a document released by the Office of the Chief Actuary (OCA) of the Social Security Administration that makes an estimate of $13 billion Social

---

[7] Institute for Taxation and Economic Reform: www.itep.org/pdf/undocumentedtaxes.pdf.
[8] CBO report: www.cbo.gov/sites/default/files/12-6-immigration.pdf.
[9] US Census: www.census.gov.
[10] Thanks to David Kallick for a helpful conversation on these issues.

Security revenue from the undocumented in 2010.[11] According to the Bureau of Labor Statistics Consumer Price Index calculator,[12] this equals about $14.2 billion in 2015. (I searched for a conversion calculator and found a commercial site that cited the BLS and returned the exact same result, confirming the more basic source.) I also learned from my email correspondence that people in the business would love better data on federal taxes on immigrants.

What about other federal taxes besides Social Security? At first we found no information but argued that the undocumented population has a lower estimated median income than the documented – low enough that the bulk of the taxes would be for Social Security. Later we found out how undocumented persons can pay taxes through an Individual Taxpayer Identification Number (ITIN), separate from Social Security. The US Internal Revenue Service (IRS) reported in 2010 receipts of $870 million through ITINs, or $0.95 billion in 2015 dollars – consistent with the thought that it would be much less than Social Security payments. Note: we couldn't find a primary source for this figure, which first appeared on a CNBC site that quoted the National Immigration Law Council, an immigrant-rights group. We then found the number in a memo of the Treasury but couldn't locate the original IRS report.[13] There is some controversy around the figure since some ITIN users were found to claim the additional child tax credit, to the tune of $4.2 billion – but does the credit benefit the worker or the additional US-citizen child? And is a lesser tax paid not a tax still paid? Ultimately, we cite the number with an asterisk.

Collecting the data we have, to two significant digits, gives the following assessment.

Type of revenue	Amount (in billions of US dollars)	Source
Social Security payments	14.	OCA
other federal income tax	0.95	IRS*
sales taxes	7.1	ITEP
property taxes	3.6	ITEP
state income	1.1	ITEP
secondary effects	???	–

Further comments. The Suspense File of the Social Security Administration holds receipts from payments that do not match with known Social Security holders. In 2003, the Suspense File received $7.2 billion in payroll receipts.[14] The Office

---

[11] www.socialsecurity.gov/oact/NOTES/pdf_notes/note151.pdf.

[12] BLS Calculator: www.bls.gov/data/inflation_calculator.htm.

[13] www.nilc.org/issues/taxes/itinfaq/,
www.treasury.gov/tigta/auditreports/2011reports/201141061fr.html#_ftnref13.

[14] Statement of James B. Lockhart III, Deputy Commissioner of Social Security, Testimony before the House Committee on Ways and Means, Subcommittee on Social Security, Subcommittee on Oversight, Hearing on Strengthening Employer Wage Reporting, February 16, 2006: www.ssa.gov/legislation/testimony_021606.html.

of the Chief Actuary states that of the figure of $13 billion in 2010, some of the receipts go into the Suspense File, while others are credited inappropriately to other taxpayers – so these numbers from different sources have some consistency.

> In total, the receipts from undocumented workers are estimated at $27 billion plus ???.

The question marks stand for secondary effects, which, as we now discuss, might turn out to be quite large.

Secondary Effects

We have mentioned some of the secondary benefits from the labor of undocumented workers: their work earns money for their employers, who pay taxes on that money; they buy things, and the increased demand on goods and services means that jobs must be created to fill these needs; the money they spend is income for someone else, who will pay taxes on it.

There are secondary effects that are negative, as well. A common argument of the opposition to the undocumented workforce is that it leads to job losses for authorized workers. If the undocumented were not present in the first place, those jobs might be filled, leading to lower overall unemployment (which is both an economic and a social issue). If undocumented workers work for less money than authorized workers, then there may be a net decrease in payments and taxes. These deficits would be mitigated if the two pools of workers did not compete for the same jobs.

In any case, quantifying these effects is terribly difficult, and this is the main reason we will only draw cautious conclusions from our study. (That doesn't mean there will be nothing learned!) Nevertheless, let us try to get a sense of how much of our national domestic product the undocumented account for. A story on National Public Radio in 2006 quotes a "recent" Pew Hispanic Center report (evidently, no longer viewable on its website) as saying the following:[15] about 5% of the overall workforce is undocumented; about a fifth of undocumented immigrants work in construction, and this labor comprises about 17% of the construction workforce (in the story, anecdotal evidence is shown that the proportion is even higher). Albeit old, imperfect, and partial data, we might try to attach a crude dollar amount to the portion of the economy that a fifth of the undocumented account for. What is the value of 17% of the construction industry? The Bureau of Economic Analysis reports that construction contributed to the Gross Domestic Product (GDP) an amount of $1,217 billion in 2014.[16] If we multiply this by 17%, to approximate the share from undocumented labor, we get about $200 billion, and if this represents the contribution of a fifth of the undocumented, then you get a figure of around *$1 trillion*. (The GDP is about 17 trillion.)

---

[15] www.npr.org/templates/story/story.php?storyId=5250150.
[16] www.bea.gov/industry/gdpbyind_data.htm

This calculation rests on the further assumption that the undocumented labor force contributes equally, on average, to the documented labor force. The assumption is most likely false, so don't go citing it! The Pew Center is reported to say that very few undocumented immigrants hold white-collar jobs. So we should be careful to point out that we are measuring the economic output of the sector, not the income of the particular labor force or whether the jobs it represents can be otherwise filled. Also, we don't know that the undocumented labor force that does *not* work in construction is similarly productive, as we assumed when we treated this fifth as representative. Whatever the precise figures, it is clear from this back-of-the-envelope calculation that the undocumented workforce contribute a huge absolute amount to the overall economy.

There is another way to think about the secondary benefits. According to Pew, there are about 12 million undocumented immigrants. The household income is in the $30,000 range. If we assume four per household (this is a rough figure, so the details are not important – and many households have people with a mix of statuses), this means a gross income of about $100 billion. Much of that money will be spent, meaning added income for others and added tax revenue in the US.

The size of these figures – the contribution to the GDP and the overall amount of wealth – means that the ??? might be very large, even dominating whatever figure we ultimately deduce from known areas.

Expenses

We now collect data on education, health care, and law enforcement, then use them to estimate costs.

An important distinction to note is the difference between federal, state, and local budgets, since all three weigh into providing educational services, health services, and law enforcement to the US populace. It may be a good idea to begin with a rough overview of spending. On the state level, according to the Center on Budget and Policy Priorities,[17]

- the states spent a total of $1.1 trillion in 2013
- 25% of these expenses went toward K–12 education
- 16% of these expenses went toward health care (e.g., Medicaid)
- 13% of these expenses went toward higher education
- 4% of these expenses went toward corrections

We now turn to the separate lines of expense.

Remark 5.2 The following data, imperfect as they are, are collated from hundreds of Web searches, click-throughs, and reviews of documents. There are data for this,

---

[17] Center on Budget and Policy Priorities: www.cbpp.org/research/policy-basics-where-do-our-state-tax-dollars-go.

that, or the other thing, and it's often not the data you want or from a source you can trust. The process was tedious and messy. We would begin our search with general terms and see data, often from commercial or advocacy sites, then hone in on the source of the data. We would eventually learn the most relevant government agency, but navigating websites can take time and may prove fruitless. Furthermore, our country has many state governments, so we were pleased when one think tank or another had already gone through the process of aggregating state data. The exercise was instructive: the next time we're looking for something, we'll probably go straight to the CBO or some such proven source. ▲

Education

In addressing education, we will try to assess the educational expenses of children who are themselves without documentation, noting that many children of undocumented residents are citizens, as anyone born in the US is a citizen at birth.

The Pew Hispanic Center[18] provides us with an approximate number of undocumented immigrants in the ten states where they reside most, and the Census[19] provides us with a per-pupil amount of state spending on K–12 education.

Now there were about 2 million undocumented school-age children (5–17 years old) in the US in 2006 (CBO, Pew), when Pew estimated 11.9 million undocumented immigrants living in the US. So the fraction $2/11.9 \approx 17\%$ of the undocumented population were of school age. More recently, Pew estimates in 2012 about 11.2 million undocumented immigrants living in the US, as well as roughly unchanging demographic trends,[20] so we take these numbers as current for 2015. Pew updates its numbers every three years, so these are the best numbers for 2013, the year for which we have spending data.

A plan: we know how much each state spends on education, and we have estimates of the undocumented populations of various states. So if we know what fraction of them are school age ($\approx 17\%$), we can estimate how much is spent on education for undocumented residents in these states.

Now while the fraction of school-age undocumented residents may be different from state to state, we will *assume* it is roughly constant so that we can use it with the Pew 2012 figures to estimate the undocumented school-age population by state (namely, 17% of those figures). This allows us to estimate the total spending on education for the ten states (the census amount quoted below includes money from federal sources). Again, assumption is needed to get us from the data we have to the data we want.

---

[18] Data taken from the Pew Hispanic Center website; see www.pewhispanic.org/2009/04/14/a-portrait-of-unauthorized-immigrants-in-the-united-states/.
[19] www2.census.gov/govs/school/13f33pub.pdf.
[20] www.pewhispanic.org/2014/11/18/appendix-a-additional-tables-4/.

State	Undoc. (thousands)	$\times \frac{2}{11.9} \approx 17\%$	Per-Pupil spent	Expenses ($ billions)
CA	2450	412	9220	3.80
TX	1650	277	8299	2.30
FL	925	155	8433	1.31
NY	750	126	19818	2.50
NJ	525	88	17572	1.55
IL	475	80	12288	0.981
GA	400	67	9009	0.606
NC	350	59	8390	0.494
AZ	300	50	7208	0.363
VA	275	46	10960	0.507

The total figure for the ten states is $14.4 billion, but we want to account for *all* the states. Adding up the undocumented populations, we see that these states account for 8.1 million of the 11.9 million undocumented, or about 68%. So *assuming* that the rate of spending on education in the remaining states is similar, this figure should represent about 68% of the national spending on education for undocumented immigrants. We compute $14.4/0.68 \approx$ $21 billion in 2013. Converting this figure to 2015 dollars still gives $21 billion, to this degree of precision.

Health Care

Sources agree that the undocumented population receives federal funding through Medicaid for emergency services, which hospitals are required to offer regardless of residency status. Undocumented residents are not eligible for other federal health care assistance. Our focus therefore was to try to find the amount spent by Medicaid on the undocumented.

The best single datum we found was from a fiscal year 2011 financial report from Medicaid,[21] which gave a precise national figure for "Emergency Services for Undocumented Aliens of" $2.17 billion. This included state and federal shares. It is possible that, as with Social Security (see below), some undocumented immigrants are receiving health care benefits fraudulently. In the other direction, much of this emergency care is for childbirths. As the newborns are documented citizens, one could argue that this money is not directed toward the undocumented (or at least not all of it). We won't try to delve further, but use this figure and note the potential for error.

Another piece of information came from the CBO report of 2006, which estimated that in a typical state, such as Oklahoma (which the CBO uses as a representative example), undocumented immigrants account for less than 1% of Medicaid

---

[21] The Medicaid Budget and Expenditure System/State Children's Health Insurance Program Budget and Expenditure System (MBES/CBES) fiscal management report from 2011 can, as of the time of writing, be found at www.medicaid.gov/medicaid/financing-and-reimbursement/state-expenditure-reporting/expenditure-reports/index.html. See in particular line 10,659, Column B.

spending. The CBO remarks that this figure is expected to be higher in some border states. The Medicaid report (line 10,681 – yes there are more than 10,000 lines in the spreadsheet!) indicates $407 billion in net Medicaid expenses. One percent would be about $4 billion, so the figure of about $2 billion is indeed less than 1%, lending some further credibility to the figure. In 2015 dollars, $2.17 billion translates to $2.3 billion.[22]

Law Enforcement

The CBO makes the following estimate of expenses for law enforcement:

- From 2000 to 2006, the Department of Justice awarded approximately $2.8 billion in State Criminal Alien Assistance Program (SCAAP) funds.

SCAAP reimburses states for expenses for the salary costs of correctional officers for undocumented criminals. Note this is not the entirety of law enforcement expenses.

Looking further into this government figure, I found a 2013 document by the National Conference of State Legislatures[23] which states that funding levels of the SCAAP in 2011 were $273 million per year (consistent with the above figure of about $2.8B/7 = $400M/yr, since the document remarks that funding levels had dropped since the time of the CBO report), and that this amount represents "about 23 percent of total costs submitted by state and local governments." If we take this figure as fact, and assume all submissions are justified, this would put costs at about $273M/23% = $1.2 billion, or $1.3 billion in 2015 dollars.

We have neither identified nor estimated other law-enforcement costs related to the undocumented population, such as the costs of patrolling or investigating criminal behavior. In not including them, we are making an implicit assumption that they are relatively small compared to the figures cited, or that patrolling expenses would have been the same for the communities patrolled with or without the undocumented among the population. This is an error in our analysis, hopefully not a significant one.

Other Costs

As for other expenses, the report of the Social Security Administration estimates $1 billion in payments to unauthorized recipients.[24] That report itself rested on assumptions that about one-fourth of undocumented residents in the eligible age range have a Social Security number, and that their benefits amount to half the average payout. In 2015 dollars, this accounts for $1.5 billion.

---

[22] On the higher end, an organization opposed to immigration, the Center for Immigration Studies, estimated the cost at $4.75 billion per year. This number is derived indirectly as a percentage of total expenditures for the uninsured. Since Medicaid has a line item for the precise figure we want, we use the lower number. Even still, as we shall see, the difference of a couple of billion dollars will not affect our eventual conclusions.

[23] The NCSL SCAAP data are found on the following website: www.ncsl.org/research/immigration/state-criminal-alien-assistance-program.aspx.

[24] www.socialsecurity.gov/oact/NOTES/pdf_notes/note151.pdf.

We cannot account for other costs or secondary effects.

> Our estimate of costs, in billions of dollars, is $21 + $2.3 + $1.3 + $1.5 + ??? $\approx$ $26 + ??? billion.

Adding It Up

We estimated the economic benefits from undocumented immigrants as $27 B + ???, where ??? represents secondary contributions. We estimate the costs as $26 B + ???, where here ??? represents secondary or unknown costs, a different number. This gives our final tally:

> a net revenue benefit equal to about $1 billion + ???

where ???, which may be positive or negative, is all the unknowns and downstream effects together.

We have seen that the unknown secondary benefits could easily be as large as the ones that were counted. And there were potential negative secondary effects, as well. In the end, the fact that our uncertainty may swamp the value we computed means that we cannot state a definitive number – we can't even tell if it would be positive or negative!

Anyway, there were many assumptions and estimates in our numbers, such as that the fraction of undocumented immigrants of school age remained constant from 2006 and did not vary significantly from state to state. Another rough estimate was the amount of Medicaid services received fraudulently. There may be other factors we completely overlooked. The point is that our figure of $1 billion has an error bar of several billion dollars at the least. How big are these numbers? Total tax revenues for the US in 2015 were $3.25 trillion, so a few billion represents significantly less than 1%.[25] In the next section, we will argue that, nevertheless, our finding is actually not inconclusive.

Remark 5.3 Numbers help to prove a point, but they can never tell the whole story.

Uncle Tom's Cabin, by Harriet Beecher Stowe, is an antislavery novel published in 1852 during a period of national conversation about slavery and abolitionism. Despite the scholarly debates throughout the country, it was Stowe's novel that resonated with Americans. The runaway best seller helped sway public opinion by putting a human face on the subject of slavery.

The history of this book reminds us that all of our arguments, always, must be consistent with the cultivation of human dignity.   ▲

---

[25] www.govinfo.gov/content/pkg/BUDGET-2017-BUD/pdf/BUDGET-2017-BUD-9.pdf.

Conclusions

Potential errors from uncertain data and assumptions we needed to make were so large that we can't even say with certainty if we found a net positive or negative effect. So what's the point? What does our analysis show?

One conclusion we can safely make is that our analysis of tax receipts and known expenses shows no known huge impact on revenues, either way. This may already lead us to preference other, noneconomic grounds for deciding policy issues related to undocumented immigrants.

All this is not to say there is not a large effect on the economy as a whole. The unknown secondary benefits to the economy (whether or not measurable by taxes and known expenses) might completely swamp the analysis we have performed.

Remark 5.4 The Streetlight Effect refers to a classic joke. A person is searching for keys in the grass under the glow of a streetlight. A police officer comes by and helps look. After some time searching: "Are you sure you lost them here?" "No, I lost them over there." "Then why are you searching here?" "This is where the light is!"

Lesson for us: just because we were able to find data for certain aspects of the immigration equation doesn't mean that those are the most relevant parts.   ▲

Even if the net unknowns of the downstream benefits or hidden costs turned out to be small or negative, that would not necessarily be economic justification (setting aside the moral and societal considerations) for a massive deportation effort. Dismantling the current system with its millions of undocumented employees could be terribly costly and destabilizing. Many historical examples show that seemingly minor disruptions to the economy can have major repercussions.[26] We must be careful not to overstate the importance of our exercise. Trying to decide even just the economic question on the basis of a single year's tax revenue is shortsighted. Long-term effects might dominate the economic impact. We're going to be here for a while, so discounting these effects is not necessarily responsible, even if modeling them precisely is nearly impossible.

That said, suppose we had determined with great certainty that there was a net cost of about 1% of total revenues – then what? Some would argue that this demands that we tighten our national borders and prevent further losses. Some would argue that this means we should issue more work permits so that more revenue is collected, thereby strengthening our economy. An analysis is not necessarily a policy recommendation.

Finally, we must once again stress the point that economics itself is only one piece of a major policy decision. Ethics, community, culture, and tradition are all part of the equation – and *even more difficult to quantify*! In fact, our conclusion that direct economic costs were relatively minor compared to the scale of the national

---

[26] This falls under the Law of Unintended Consequences.

economy might suggest that these other considerations are the ones on which to steady our focus.

Our analysis was not sharp. Even if it were, there would still be plenty of room for debate. People are people, and they'll have their opinions. We can't change that, and don't want to. We're just here to ensure that those opinions are supported by responsible arguments.

Summary

We analyzed the economic impact of the undocumented population in the US by looking at tax revenue versus government expenditures. Admitting that this was a narrow lens to look through, we decided that the information would nevertheless be illuminating. Our sometimes sprawling discussion indicated the can of worms that this complex question opens up but also allowed for a nuanced exploration of bias and reliable sources.

We had to make assumptions to create approximate figures, e.g., estimating downstream benefits through looking at the construction industry or extrapolating data from ten states to the nation as a whole. Our assumptions and incomplete data – and imperfect understanding of the totality of economic contributions – created errors as large as the figures obtained. Still, since the known costs and benefits were about equal, we concluded that one could not affirm any large direct impact on revenues, though there were some indications of large positive secondary impacts.

Our inconclusive search suggests the relative importance of other considerations, such as geostability and societal and humanitarian perspectives, against a single-year, revenue-based analysis.

Exercises

1 Referring to the "Tactics in Argumentation," come up with one short (flawed) argument for each of the six different tactics (so six total arguments).

2 Explain how different studies on undocumented workers might arrive at opposite conclusions. Name five factors that might contribute to the discrepancy.

3 Pick two countries and find out how their immigrant populations (documented or not) have changed over the past decade. Try to pick two countries that are trending in opposite directions. (This exercise involves research.)

4 I read in the news that last winter was the warmest on record, and yet I and most of my friends thought it cold as ever. What could be some sources of discrepancy between my impression and the reported warming trend?

5 A news story states that the average temperature of the Earth is rising. What could the reporter mean by "average temperature"? How do you think

someone arrives at such a quantity? Suppose you had all the resources you need. How would you compute that average temperature *today*? Do you think the "average temperature" is measured for a single day? Why or why not?

(This issue is explored in some more detail in Problem 4 of the Chapter 8 Exercises.)

6 Suppose undocumented workers in the US make on average 72% of the taxable income that documented workers make, and suppose undocumented workers represent 5% of the workforce, and suppose that the overall US revenue from individual income tax is 1.4781×10^{12} (these numbers are approximately true in 2015). Now suppose (this statement is *not* true) that undocumented workers pay no taxes, so that all Treasury receipts are from documented workers. What would be the total value of income tax that undocumented workers would have paid on all of their earnings? You may assume that the tax rate is the same for all taxpayers.

How could this figure be used in arguments by both immigration advocates and opponents?

7 In 2018, the Philadelphia Eagles won Super Bowl LII, the US football championship. The city celebrated its first football victory since 1960 by throwing a parade. Fans flooded the 5-mile-long parade route, standing side by side (two people per meter) and on average 20 people deep on both sides. Estimate the number of people in the crowd.

A journalist (Tom Avril) reported on the work of crowd-size experts who did a similar calculation (in much greater detail) and was flooded with comments from angry Eagles fans who had bandied about much higher numbers. Is there any harm in going with the larger, emotional figure cited by a couple million fans? Can you think of why it might be critically important to have accurate estimates?

8 Find a patently biased source on the immigration issue (whether or not you agree with the bias). Now state a specific argument proffered where quantitative reasoning is misapplied, and explain how. (This exercise involves research.)

9 You are tasked with estimating the economic benefit to the California economy of undocumented workers in the fruit- and nut-growing industry.
   a Which pieces of data will you need to reach a conclusion?
   b Find these pieces of data. If a number you need is not available, try to figure out a way around it by using other numbers to estimate the desired quantity.
   c Use your data to estimate the economic benefit.
   d What are some possible improvements to your model?

10 Think of a topic for a project related to this question.

Projects

A Should the city of Chicago raise its minimum wage?[27]
Be sure to explain how you will be interpreting this question, and discuss
the benefits and drawbacks of your interpretation. Formulate a *quantitative*
approach to the question. Describe your approach to the data, including a
discussion of the sufficiency (or not) and reliability (or not) of the data you
obtained, as well as the limitations of your analysis.

B Should the US government increase or decrease its current level of spending,
as a fraction of the total budget, on _____? You may pick whatever
group or topic of interest you like to place in the blank, even a controversial
one. Just be sensitive and respectful.

C The figure of "12 million" undocumented residents in the US was used several
times in this chapter. Where does this figure come from, and how accurate is
it? Be sure to discuss the methodology used in constructing this figure.

D Explain the use of numbers in several civil rights struggles. Examples might
be patterns of discrimination in hiring, housing, law enforcement, or election
workers.

---

[27] Since the writing of this book, Cook County passed a law raising the minimum wage in stages
over time. You may nevertheless consider the question fresh, or choose another city.

6

Should I Buy Health Insurance?

Appendix Skills: Probability

OVERVIEW

In this chapter, we'll do the following.

1. *We consider the question of insurance: how does it work, and how do we decide if we need it?*
2. *We learn how to evaluate insurance fees with expected values, by estimating risks and costs.*
3. *Next we encounter the usual problem: we can't find perfect data. So we must assume the person in question falls into a cohort for which we have information.*
4. *We then use hospitalization rates and costs to estimate health care costs with no insurance.*
5. *Finally, we compare estimated expenses with fees on the available health insurance plans to determine which choice of plan (or no plan) would be most cost effective.*

What Is Insurance, Anyway?

Before diving in, let us perform the usual task of asking how we are going to interpret and answer the question. We're discussing the value of insurance. Generally speaking, is insurance worth it for the average person?

> *What do you think?*
>
> *Think of your response and jot it down.*

In some sense, the answer is no. Insurers of all kinds take in money from lots of policy holders and pay it out to those whose circumstances require it (an accident, an illness, a fire), as stipulated in the fine print. Insurers are not philanthropists: they make money, so on average their clients must "lose money" when they buy insurance.

> *Then why do they do it?*
>
> *Think of your response and jot it down.*

They do it because there is a small chance that a catastrophic event will be too costly to bear without help. If there's even a tiny chance that we'll lose our home in a fire, most of us would be willing to pay for the peace of mind that we won't also "lose our shirt." That said, some live more carefree than others, especially when it comes to our bodies. We who think we're invincible wouldn't pay out a dime for health insurance, while the cautious among us believe the security is worth the cost.

The best way to approach such a question, then, is to estimate the costs with and without insurance. These costs will depend on our behavior and approach to spending on our health, so the answer will inevitably be "it depends." Nevertheless, we will try to assess these costs quantitatively so that we are making a conscious decision based on knowledge rather than a gut choice based on instinct.

What, Me Worry?

So this chapter is about insurance. We will begin with an extended analysis of a simple, but illustrative, example. I recently rented a car, and it came with insurance, but there was a $2,000 deductible.

What does this mean?

Think of your response and jot it down.

Insurance with a deductible of $2,000 means that while the insurer will pay for the damage caused by an accident, the first $2,000 comes out of your pocket – so if the damage is less than that, you get nothing.

So this car rental company calls me up on the phone. I had already had a reservation for a car that came with some insurance, but the woman on the phone asks me if I want to purchase the "Stress Free" insurance policy,[1] with *no* deductible, for just $100.

Should I take the "no worries" policy?

Think of your response and jot it down.

Different people will no doubt answer differently.

If you gave pause while answering, why?

Think of your response and jot it down.

Even more importantly:

Write down the main factor to consider in answering this question.

To help think about it, consider this question:

Would you buy Stress Free if you knew you'd crash the car?

Think of your response and jot it down.

---

[1] This really happened, though I don't recall the exact cost of the "Stress Free" insurance.

Surely you would – though in fact, you shouldn't rent the car at all if you know a crash is coming! Just as surely, you wouldn't buy insurance at all if you knew for certain that there would be no accidents, fender-benders, etc. Reality lies somewhere in between these extremes, so we should conclude that the answer depends somehow on the *likelihood* of your crashing.

Remark 6.1 Note what we've done. We took a real-world example and sharpened our interpretation by considering extremes. This is a very useful method for clarifying the questions which will need to be addressed, in this case the probability of an accident. There are pitfalls to such exaggerations, too. For example, a piece of legislation regulating hydrocarbon emissions is sometimes pilloried as the start of a "slippery slope" that will lead to a "nanny state." While there may be legitimate reasons to oppose any particular bill, such hyperbolic statements are typically exercises in rhetoric rather than honest attempts to quantify the need for pollution control. (Oh, and not all slopes are slippery.) ▲

Now the likelihood of having an accident *itself* depends on different factors.

Write down a list of three factors.

There are many. An actuary is the person who works for the insurance company to set the terms and price of the policy based on all the risk factors. Some of these are the age, sex, and driving history of the driver – but these are all determined by the time you step up to the rental counter. Further factors that vary with each rental are road conditions, time of day, weather, visibility, and how much you will be driving.

In my case, I was going to rent the car to drive a few hours to a remote town on country roads, then tour around there for a few days of short trips (under an hour), and finally drive the few hours back to return the car. All told, compared to my family's usual driving rate, this would be perhaps a week or two of driving, albeit in unfamiliar conditions, but with no chance of snow and no night driving.

Note how we are teasing quantitative information from what looked like a narrative of my intended driving use: I identified how much I would be driving and under what conditions. I then made a useful comparison to the familiar.

By comparing (albeit somewhat haphazardly) my intended driving use with my driving under familiar conditions, I can now access lots of data regarding the likelihood of an accident. In well over ten years of regular driving, I have made two insurance claims (both for over the $100 price of the Stress Free option, of course – nothing is cheap when it comes to car repairs). One was under ideal conditions (scraped a car while pulling out of a parking lot), and one was in snow (scraped the side of a car at school pick-up in a street that was narrowed due to the large snowbanks). Whatever the cause, these were two accidents in at least ten years of driving – let's say 500 weeks, since the timeline is not terribly precise anyway.

So my accident rate is 2 in 500 weeks, or 1 in about 250 weeks.

What is yours?

Think of your response and jot it down.

With this information, what is the chance of my getting into an accident in two weeks of normal driving (which we have equated with the amount of driving for the rental period)?

Write down your answer.

The answer is $(1/250) \times 2 = 1/125$, which we'll take to be about one in a hundred, or 1%. (Again, none of the numbers used here are precise. One significant figure will do.) Of course, it may be that my two accidents were both flukes, or that I should have been in two more accidents but expert drivers around me avoided collisions. It's hard to say, since despite the large number of weeks, there are only two data points for accidents. An insurer would have more reliable data about accident rates for people "like me," but I can only work with the data I have about my own driving, which after all may or may not be typical for people "like me."[2]

Remark 6.2 Note that after performing our calculation, we restored a single significant figure. Also note that we are making a *conservative* estimate, meaning the probability should actually be lower than our estimate, as we rounded 125 down to 100 and the number of years of driving ("well over ten") down to ten.   ▲

Okay, now this extra insurance will cost $100 and potentially save us $2,000. In fact, though both my accidents were just scrapes, the damages were around $2,000 each time (I was unlucky enough to dent a BMW). To simplify matters, we will make an *assumption* that the cost of an accident will be at least $2,000.

With this assumption, we will be ready to attack the problem of deciding on a rental insurance option. Now before I ask you the central question, let me ask you this. We've computed the probability of an accident. What should we compute with this quantity in hand?

Think of your response and jot it down.

Here we have to go back to the original question: we are trying to decide between two options. Each one has an expected cost to us, and that cost will depend on the probability of getting into an accident. So what should we compute to resolve this question?

Think of your response and jot it down.

---

[2] An insurer may interpret "like me" to mean the same sex, age range, income bracket, type of residence, and type of vehicle, whereas I might interpret "like me" to mean having the same cultural tastes and political bent. What's most relevant is the extent to which the traits we use to characterize individuals bear on their driving patterns. No doubt the insurers are better at judging these things than I.

We must compute the expected cost to us with insurance and without, then compare. (See Appendix 7.1 for more on expected values.)

Without insurance, our expected cost is the $2,000 price of an accident, which comes with probability 1%, and the cost of $0 for no accident, which comes with probability 99%. So the expected cost is the weighted sum

$$\$2,000 \times 0.01 + \$0 \times 0.99 = \$20.$$

The cost to us with the Stress Free insurance is $100 (whether or not we get into an accident). Now since $20 is less, we should decline the Stress Free option.

Remark 6.3 Based on the reasoning above, I declined the Stress Free additional coverage, and sure enough the trip was not stress free! Just about an hour into the long drive, I noticed a tiny chip in the windshield. I knew that these chips were easily fixed, but I was in no position to fix it. I was frustrated and nervous about that chip throughout the trip. When I returned the car, they didn't even look it over, but I read in the contract that they could levy a charge to my credit card even after I returned it. I continued to worry about it and called up my credit card company to see what my rights would be if they decided to charge me for the windshield repair. In the end, they didn't, so while I saved money, I did it at the cost of worry. To properly revisit the question, I should try to put a monetary figure on the amount of worry I might incur even if there were no accident. If I had to do it again, I'd take the insurance! ▲

Clearly, there is a magical probability beyond which insurance pays off. Let's try to find it. It is unknown to us, so we will need to work with variables.

> Tip: when you need to run an argument without knowing the specifics, use a variable to indicate the unknown quantity with a letter. If the letter scares you, first run the argument with the number 3, but don't ever multiply out or cancel the 3 in your calculations! Afterward, put a line in front of the 3 to make it look like a B: you've just used a letter! Now if you're not fond of B, replace it by the letter of your choice!

Let's call p the probability of getting into an accident. We will calculate the expected costs with and without insurance *as a function of p*, then compare. The magical tipping point will be the value of p at which the two costs are equal.

In all cases, the cost with Stress Free insurance is $100. Without the Stress Free option, the cost of an accident is $2000, and this occurs with probability p, giving an expected cost of $\$2,000p$. Setting $\$2,000p = \$100$ gives $p = 1/20$. Since I am estimating the amount of driving as being equivalent to about two weeks' worth of normal driving, this means that as long as there is less than a one-in-twenty chance that you don't get into an accident in two weeks, you should be okay (on purely monetary grounds!) to decline the additional coverage. Or equivalently, since 1/20 per 2 weeks equals 1 per 40 weeks (recall "per" is "divided by"), if you get into an accident less frequently than once every 40 weeks, you might want to decline.

Remark 6.4 The part about worrying over the windshield is real. Some people are risk averse, meaning they don't want a chance of a large payment even if it means that on average they will pay more. That's how insurers make money – on average, the customer "loses" – but most of us are okay with that, as long as it means, for example, that there is no chance that we'll lose the full value of our home from a fire, flood, or whatnot.   ▲

Health Matters

Okay, we've tackled the "easy" problem of a simple decision on car insurance, but the driving question of this chapter (pun intended) is about purchasing *health* insurance. This is a considerably more complicated question.

> *Why?*
>
> *Think of your response and jot it down.*

The main reason is that there are many more events and factors to consider. Still, the question is analogous.

> *What here is the analogue of crashing the car?*
>
> *Think of your response and jot it down.*

Instead of insuring against a car accident, you are considering insurance against getting sick, or more precisely needing medication or the services of a physician. Your likelihood of needing such care depends on your age, sex, medical history, family history, genetics, lifestyle, diet, exercise regimen, job, environmental factors, and many other variables. On top of that, the range of expenses varies tremendously, from somewhere around a hundred dollars to hundreds of thousands of dollars, depending on what medical condition arises – and the different conditions have their own probabilities.

So far, we have considered a single cost ($2,000) with a single probability (p) and performed an estimation of the expected cost to us ($2,000p$). (The other cost, $0, with probability $1-p$, did not affect the calculation.) Before contemplating multiple costs and probabilities, we will warm up with two- and three-cost scenarios.

How about this? Your friend is throwing a dessert party and has baked cookies but suddenly realizes that the guests are coming soon and there's no milk! Your friend asks you to drive to the store to pick up a gallon of milk. At the intersection, there's a traffic light. You decide that in the interest of time, if it's green, you'll go straight to the bargain grocery store, where you can get a gallon for $2.50, while if it's red, you'll go to the gourmet store located just at the intersection, where it costs $5.75. The traffic light preferences the road you are on, so there is a 60% chance that it will be green and a 40% chance it will be red.

> *What is the expected cost to you?*
>
> *Think of your response and jot it down.*

A simple way to do this without formulas is to imagine going to the store ten times. On average, the light will be green six times (for a cost of $6 \times \$2.50 = \$15.00$) and red four times (for a cost of $4 \times \$5.75 = \$23.00$). In total, you'll spend $38.00, on ten trips, meaning on average you'll spend $3.80 per trip. This is the correct answer, but to proceed more systematically, let us recall the lesson on expected values from Appendix 7.1: we calculate the expected cost as the sum of the costs times their probabilities: $0.6 \times \$2.50 + 0.4 \times \$5.75 = \$3.80$.

Before moving on to the general case, we'll consider a new twist. Suppose now that there is also a regular convenience store if you turn right on a green arrow (no turn on red here). The convenience store sells milk for $3.75/gal, so you would go there instead of the gourmet market, but not instead of the budget grocery. Suppose that the chance of green is still 60%, the chance of red with no green arrow is 30%, and the chance of a red with a green arrow is 10%. So now there are three distinct possibilities: green (probability 0.6; cost: $2.50), red with no green arrow (probability 0.3; cost: $5.75), and red with a green right arrow (probability 0.1; cost: $3.75). The expected cost is therefore

$$0.6 \times \$2.50 + 0.3 \times \$5.75 + 0.1 \times \$3.75 = \$3.60$$

(remember, order of operations says we do the multiplication before any addition). The expected cost went down because the possibility of a red with green right arrow mitigated the expense when the light was not green.

How to consider a general scenario where any number of events can occur with various costs and associated probabilities? We needn't have a definite scenario in mind or a even a name for the occurrences or their costs and probabilities. But we have to label them *somehow*, so again, we use *variables*. If the events $E_1, \ldots, E_N$ have costs $C_1, \ldots, C_N$ and occur with probabilities $P_1, \ldots, P_N$, then the expected cost is

(6.1) $$P_1 C_1 + P_2 C_2 + \cdots + P_N C_N.$$

Now we know how to answer the question about health insurance. We need to itemize all the possible ways you can require medical care and tally their probabilities and costs, then compute the expected cost to you. Most health insurance plans have a one-year enrollment plan, so we will compute costs and probabilities for one year out and compare the expected costs with the health plan under consideration. In fact, we would need to complete this whole exercise for each plan, since the out-of-pocket costs for each event depend on the plan under discussion. (A spreadsheet would come in handy.)

Remark 6.5 It may be foolish to approach your health spending from the basis of minimizing costs rather than maximizing health, or even looking just one year ahead rather than taking in the long-term picture, but we have chosen (perhaps carelessly) to frame the question in this way. Modeling expected health outcomes could prove difficult, and surely long-term modeling of medical expenses can get dicey; our choice is primarily one of convenience.   ▲

With this understanding, we have proposed a mathematical model and now need to do our research. This is where the real-life problem gets tricky. National insurance companies have access to scads of data, and employ teams of actuaries to analyze them. You the individual are at a competitive disadvantage in making the choice whether to insure. On your side, however, is the choice among different insurers.

Mathematically, there is also the difficulty that not all the probabilities for health expenses are independent. For example, if you are hospitalized for asthma once, there is a better chance that you will require hospitalization again at a later date. This is different from flipping fair coins, where the occurrence of a head once has no impact on the next or future tosses. For simplicity, we will not build such a sophisticated model – yet another example of how we are engaging in shoddy practices compared to a professional![3] Surely no insurer would hire us as an actuary at this point. But creating an introductory model based on some explicitly stated assumptions, even if they are unrealistic, is a good way of understanding what the professionals do and arming yourself against salespeople and advertisers. It is also a starting point for building a more sophisticated and realistic approach in a future iteration.

Lana's Choice

For this model, let us consider the case of a female in her thirties living in Connecticut in 2015. Let's call her Lana. We might have considered other possibilities, or taken an average person from across the US, but, truth be told, we were only able to find excellent data for the state of Connecticut. Thanks, Gov'nuh!

Our first and main task is to calculate our expected medical expenses without health insurance. (Calculating expenses with insurance is more straightforward.) To do so, we need to find the probability of an "accident" occurring as well as the cost of such an event. To simplify the calculations, we first consider only one event – being hospitalized for any reason – instead of finding the individual costs and probabilities of contracting several different illnesses. Later, we will make some rather crude corrections to account for other health expenditures.

To find the probability of being hospitalized, we need to know how many people were hospitalized and the total number of people in the group we are considering. The probability of hospitalization for our cohort will then be the first number divided by the second:

$$P(\text{being hospitalized}) = \frac{\text{number in cohort who were hospitalized}}{\text{total number in cohort}}$$

---

[3] A more thorough analysis would require conditional probabilities, at the least. See Appendix 7.2.

Even this is a bit of a simplification, as the data counted hospitalizations, and that could include multiple hospitalizations for the same person. We will assume that the overall effect of this on the probability is small. (This is akin to assuming that the hospitalizations were independent events.)

First, we estimate how many females in their thirties are hospitalized in a given year. We will use data from the 25–44 age range because that is the closest information available to us. For this and other estimations of women in their thirties, we are forced by our data to engage in another assumption: that there are no significant hospitalization or cost spikes (or troughs) differentiating women within the age range of 25–44 yet outside their thirties. This assumption is justified by looking at the overall scope of data: it doesn't appear to vary too rapidly with age.

Even with this assumption, we don't have quite the data we want. Here's what we have:[4] age- and sex-based data for the top ten reasons for hospitalization, as well as total number of hospitalizations by sex. We also know the total number of females, age 25–44 in Connecticut in the same year (2012).[5]

> *How could we estimate the probability we want?*
>
> *Think of your response and jot it down.*

To answer the question, since we already have the denominator (total number in cohort), we need to estimate the number of total hospitalizations in the cohort. We can get this as follows: we have the number of top-ten hospitalizations in Lana's cohort, so if we knew what fraction of *all* hospitalizations in this cohort are top-ten, we'd be set. But we do know what fraction of all hospitalizations are top-ten across all cohorts, so we will **assume** that this fraction applies to Lana and deduce the number of total hospitalizations in our cohort that way.[6]

Here are the numbers. Out of 172,117 total females hospitalized in Connecticut, in 2012, 139,352 were hospitalized for a top-ten condition, or roughly 80%, 19,047 females in the 25–44 age range were hospitalized for a top-ten condition. *Assuming* this represents 80% of their total hospitalizations, we can estimate the total hospitalized Connecticut females in their thirties at about 24,000.

The female population of Connecticut between the ages of 25 and 44 in 2012 was approximately 450,000. This means for this year that this demographic has a chance of hospitalization of[7]

---

[4] https://portal.ct.gov/DPH/Health-Information-Systems--Reporting/Population/Annual-State--County-Population-with-Demographics.

[5] www.ct.gov/dph/cwp/view.asp?a=3132&tq=388152.

[6] This is similar to the kind of assumption we made in Chapter 5 when we needed to extrapolate education data from the ten states with large populations of undocumented residents to the country as a whole.

[7] We have unnecessarily introduced a slight error for ease of exposition. We have kept just one significant digit here so that we may write the probability as a simple fraction. This way, it will be more clear how we are using probabilities and that 19/20 is the probability of *not* being hospitalized.

$$\frac{24{,}000}{450{,}000} \approx 0.05 = \frac{1}{20}.$$

The mean charge for a hospitalization in Connecticut was $39,000 in the year 2012. We should try to estimate this figure in 2015, adjusting for rising health costs. Rather than just guess, we search for helpful data. The Centers for Disease Control (CDC) figures from different years indicate about a 3.8% rise in health costs.

> *Assuming this trend continues, estimate the cost of hospitalization in 2015.*
>
> *Think of your response and jot it down.*

Each year the cost goes up by a factor of 1.038, so in three years we estimate the cost as $39,000 × (1.038)^3 \approx $43,600.[8]

This puts the expected hospitalization cost of a Connecticut woman in her 30s at

$$\frac{1}{20} \cdot \$43{,}600 = \$2{,}180.$$

Now we confront the fact that hospitalization is but one kind of medical expense. What about other health expenses and the probabilities of incurring them? Though we have the mathematical ammunition to address a host of probabilities and costs – the beautiful expected-value formula we derived in the previous section – we simply don't have the data necessary to deploy it. This is a common shortcoming of research in any field, and in this particular case, data on the uninsured are simply too hard to get.

Nevertheless, having arrived at a decent estimate of the mean hospitalization cost, we could tease out an estimate of the total health expenditures if we knew what percentage of health costs were due to hospitalizations. In fact, overall data are available: hospital care accounted for 38% of total personal health care costs in the US in 2012.[9]

The above datum includes mostly insured individuals. Here's where individual behavior comes into the equation:

- If we assume that the expected hospitalization cost of $2,180 for a woman in her thirties in Connecticut represents 38% of her yearly health costs – in line with the insured – then we estimate her expected yearly medical expenditures to be $2,180/0.38 \approx $5,740.
- However, this assumption may be unrealistic for the uninsured, as many without health insurance don't bother spending on their health unless a hospitalization requires it. If our uninsured woman forgoes all medical expenses outside of the hospital, then hospitalization represents 100% of her medical expenses for the year, for a total of $2,180.

  Furthermore, the uninsured might avoid inpatient hospital care entirely (due to the prohibitive costs), opting only for emergency treatment. Such a short-term

---

[8] The process of multiple fixed-fraction increases was discussed extensively in Chapter 3.
[9] www.cdc.gov/nchs/data/hus/2014/103.pdf.

decision might have long-term implications on health care costs, but we have already made the choice to analyze the question for a single year.

There is a great chasm between these estimates. It means we might have to give another one of those *it depends* answers, since the final yes or no could hinge on this difference. Or, we could try a little harder to estimate the fraction of health costs due to hospitalizations for the uninsured.

Data from 2008 (we have no choice but to use these as approximately relevant for 2015 and somewhat valid for Connecticut)[10] show that, nationwide, the uninsured were less likely to have inpatient hospital visits or overall hospital visits and use any health services overall. The percentage of uninsured who had an inpatient hospital visit was 2.9%, whereas that number was 4.6% (close to our Connecticut value of 5%) for the insured. With fewer visits, health care costs go down (at least for that year!). The ratio 2.9/4.6 is about 0.63, meaning that the uninsured can be modeled to have an expected hospitalization cost of 63% of our calculated value of $2,180, or $1,370.

According to our data, the uninsured also seemed to use other health services less than the insured, and by similar fractions: 20.6% of the uninsured had some hospital visit, compared to 29% for the insured, a ratio of 71%; 62% of the uninsured used some health service, compared to 89% for the insured, a ratio of 70%.

So how can we estimate the yearly health costs of the uninsured?

Think of your response and jot it down.

It means that we can assume that the uninsured are spending something like 63% as much as the insured across the board, giving a total of 0.63 × $5740, or about $3,600. The number would have been a bit larger if we had used 70% or 71%. We used the 63% figure because it was associated to the largest single health care expense. We could have arrived at the same number another way: if we use the new figure of $1,370 in hospital costs in place of $2,180, and still assume that this represents 38% of the total yearly medical expense, we get $1,370/0.38, or $3,600.

Finally, we must consider what the law has to say. The Affordable Care Act ("Obamacare") imposes a penalty on those who decline medical insurance.[11] For a (single, head-of-household) woman in our cohort in 2015, this amounts to the higher of[12] 2% of yearly income beyond $10,150 or $325. If we take our model person Lana to have the median income of a person in her cohort, this would mean an income of about[13] $36,700, leading to a penalty of . . .

---

[10] www.ncbi.nlm.nih.gov/books/NBK221653/table/ttt00002/?report=objectonly.

[11] This was the law when this chapter was written in 2015 and remained true for a revision in July 2017. There has been constant talk about changing it, however. In late 2017, the individual mandate was repealed. In December 2018, a federal judge ruled the whole thing unconstitutional. Stay tuned! But whatever the current law may be, the tools we develop can be used for other insurance questions and probabilistic cost–benefit analyses.

[12] www.healthcare.gov/fees-exemptions/fee-for-not-being-covered/.

[13] www.bls.gov/news.release/wkyeng.t03.htm. We averaged the median incomes in cohorts 25–34 and 35–44; the two had about the same number of people, so a weighted average was unnecessary.

What would the penalty be?

Think of your response and jot it down.

We first have to calculate 2% of the income past $10,150$. This is $0.02 \times (\$36,700 - \$10,150) = \$531$, which is greater than $325. So Lana's penalty is $531. We have to add this to her expected yearly cost of $3,600.

> Lana's total expected cost without insurance, to two significant figures, is $4,100.

Remark 6.6 It is evident how many variables the answer to our question depends on: age, gender, location, marital status, income. On top of all these, the probabilities associated to different health costs depend on a host of lifestyle variables: how much does Lana (or whoever) drink, smoke, exercise, work, drive, etc.? Much as we'd like to keep it simple, a one-size-fits-all response is plainly impossible.   ▲

Next, we need to find the expected yearly cost if Lana *does* buy health insurance. By comparing the costs, we will then be able to determine if it is more economical to go with or without insurance.

There are several different health insurance plans available to a Connecticut female in her thirties in 2015. We will compare a bronze-level (lower monthly cost, higher deductible) plan to a gold-level (higher cost, lower deductible) plan, *assuming* that Lana has four routine doctor's visits per year[14] and that in the event she is hospitalized, she will spend her entire deductible.

The bronze plan has a monthly cost of $221.72 and a deductible of $5,000. Doctor's visits cost full price up to the amount of the deductible, then are free afterward.[15] A routine doctor's visit (in the US) costs approximately $200.[16] This means that at the bronze level, the costs would be $12 \cdot \$221.71 + 4 \cdot \$200$ *plus* whatever remaining deductible would be spent on hospitalization. To calculate this, we remark that *if* Lana is hospitalized, since the price of hospitalization is higher than $4,200 (the remaining amount of deductible after her four doctor's visits, which she'd have to pay out of pocket), then there is a $1/20$ chance of paying this amount and a $19/20$ chance of paying nothing. So the expected out-of-pocket costs of hospitalization are $\frac{1}{20} \cdot \$4,200 = \$210$. Adding this gives a total expected cost from the bronze plan of about $3,700.

The gold plan has a monthly cost of $341.05 and a deductible of $750. Doctor's visits cost $15 per visit, and this cost does not apply toward the deductible. Using a similar estimation, the total cost of this plan is about $12 \cdot \$341.05 + 4 \cdot \$15 + \frac{1}{20} \cdot \$750 \approx \$4,200$.

---

[14] According to this infographic in *Forbes*, which quotes the Commonwealth Fund, Americans visit the doctor about four times per year: http://blogs-images.forbes.com/niallmccarthy/files/2014/09/20140904_Doctor_Fo.jpg.

[15] The source for these data is unfortunately now a dead link: www.ehealthinsurance.com/resource-center/ehealth-price-index. To view this and other now-defunct websites, first go to the Internet Archive's Wayback Machine at https://web.archive.org, then enter the URL there.

[16] www.bluecrossma.com/blue-iq/pdfs/TypicalCosts_89717_042709.pdf.

Based on our model, then, the bronze plan is about $500 less expensive than the gold plan. In exchange for the extra $500, the gold plan gives you the security of knowing that you won't suddenly have to outlay a very large deductible if in fact you are hospitalized. Note that a person on the bronze plan could change her habits and visit the doctor less frequently, resulting in a larger short-term savings – again, perhaps at the expense of long-term costs from undetected or untreated health problems.

Let us summarize our results.

Insurance status	Expected cost
Uninsured, no penalty	$3,600
Bronze	$3,700
Uninsured, with penalty	$4,100
Gold	$4,200

Warning! The errors produced in our analysis were probably greater than the difference between these estimates. We would need more theory to talk about error estimation, but nevertheless, it is instructive to engage in some interpretation. *According to these (suspect) numbers,* the cheapest available option in 2015 is the bronze plan, at $3,700 (not paying the penalty is not an option). Note that we can see here that the penalty has its desired effect: it incentivizes Lana to purchase health insurance. Without the penalty, she would have lower expected costs by forgoing insurance, but the penalty tips the balance.

Errors

Our analysis is *rife* with potential errors, due to the various assumptions we were forced to make, and due to insufficient data and our desire to build a rather simple model. There are other hidden assumptions, as well. For example, we ascribed the mean hospitalization cost to our woman in her thirties – yet women in their thirties may have hospitalization costs very different from the mean. Maybe they are higher due to childbirth. Maybe they are lower because women in their 30s may not need as serious treatment as the older cohorts. We don't have these data. Furthermore, our conclusions rested on an individual's spending patterns – how much of Lana's expenses will be hospital related? – and these may depend on personal choice.

Surely, our analysis can be refined. Our work may serve as a springboard to creating a version 2.0 of our model. Or at the least, it can give us a sense of the problem that might be good enough to resolve the decision.

Finally, we must recognize that, once again, we have answered the question purely on monetary, microeconomic grounds, meaning we are only considering the cost to *you* (or to Lana). Looking at the question from a broader perspective, we might ask, what is the benefit or cost to *society* of the different choices? Suppose you are a healthy, 27-year-old female and have calculated that buying health insurance – even though you might be able to afford it – does not benefit you. So you, now, might not want to get insurance. Now suppose you are the same person, but 50 years later. As a 77-year-old woman, possibly on a fixed income, who may benefit greatly from insurance, you may hope that there are more people in the insurance "pool," mitigating your expenses. The later version of you might want the present version of you to be insured.

There are further complications. When people cannot afford to pay their medical expenses, some of the remaining debt is picked up by the government. So in a sense, the "cost" these individuals pay is less than what we might calculate, but the effects of not paying can be serious, even if they are not visible to the bottom line: a bad mark in one's credit history can make it hard to get a mortgage or get loans for other purposes as well as lead to higher rates (as you become a greater risk), and can lead to difficulties in getting approved for a rental apartment or even getting a job. Also, since the government pays some of the remaining debt, one person's failure to pay places a greater financial burden on other taxpayers.

We each assess our civic responsibilities as well as our financial capabilities and do what we feel is right.

Summary

The price of health insurance is determined by the expected costs of medical treatment, essentially an exercise in probability. By finding the probability of hospitalization (at least in Connecticut, for which we were able to find good data) as well as the cost of that event, we were able to compute the expected hospitalization costs. We then inferred from these data the expected overall costs a person with and without insurance would incur, and compared. Having concrete figures helped us frame the choice of getting insurance on financial grounds. Other considerations – our anxieties or sense of responsibility – might influence our actual decision in real life.

Exercises

1 A neighborhood child asks you to purchase a lottery ticket for $5, with a chance to win $1,000. What piece of information do you need to determine whether the chance is worth taking? (That is, is it worth it from a purely monetary viewpoint? Of course you might want to support the fundraising cause.) Suppose you had that quantity you needed. What amount would it need to be to make the purchase a "wash," i.e., what's the break-even point?

2 Some credit cards come with automatic supplemental insurance when you rent a car with that credit card. Choose a particular credit card, find out their policy, and estimate the benefit to you as a potential future renter. (This exercise involves research.)

3 You have a coupon that allows you to buy milk for $1 a gallon. I would like to buy it off you. What's a fair price? Think carefully about terms and restrictions that might affect the value of this coupon.

   Hint: the answer depends on the price of a gallon of milk. Pick any reasonable price.

4 In the three-option milk-run scenario where the possibility of the red light with green right arrow is included, what is N and what are $E_1, \ldots, E_N$, as well as $C_1, \ldots, C_N$ and $P_1, \ldots, P_N$, of Equation (6.1)? What do they represent?

5   a Find the difference in maximum cost between the two health care plans discussed in this chapter. Assume four doctor's visits.
       b We found that the gold plan, with an expected cost of $4,200, will cost $500 more than the bronze plan ($3,700) on average. This $500 can be interpreted as the price of the added security you get by reducing your maximum cost, as above. Calculate the extra in expected cost that you pay for each $1 reduction in maximum cost.

          Hint: the answer is $0.18.
       c A third (silver) health care plan is available with a monthly payment of $301.77, plus $50 per doctor's visit, a cost which is not applied toward the deductible. What should the deductible be, to the nearest hundred dollars, to maintain the same extra charge per $1 reduction in maximum cost?
       d Suppose a fourth health care plan were available with no monthly payment and no charges for doctor's visits, but with a deductible for hospitalizations. What should this deductible be for the bronze plan to be the same $0.18 per dollar reduction in the maximum cost, as in Part (c) above? Is this realistic?

6 Give your best estimate of what it would cost you if your place of residence were to get robbed. Be sure to include (and find out) whether you have homeowner's insurance, renter's insurance, or any other relevant factors. (This exercise involves research.)

7 Think of a topic for a project related to this question.

Projects

A You work for a budget car insurance company that is trying to expand to home insurance. The budget line will *only* insure against fire. Your boss demands that the price of the policy be $250 per year. Expensive houses will not be insurable at this low rate, but perhaps some will. What is the highest-value house that your company can insure at this rate with no deductible? What will your deductible need to be for homes of a higher value?

B You represent a team in the National Football League seeking to sign a hot prospect out of college. How will you structure the deal – as a fixed salary, as a variable salary, or with an up-front payment followed by salary? Be sure to discuss the factors leading in each direction, as well as other factors and probabilities to consider, including the likelihood of injury. Create a specific profile for your player and your team (including salary cap constraints), and analyze the decision based on that profile.

7

Can We Recycle Pollution?

Appendix Skills: Algebra

OVERVIEW

In this chapter, we'll do the following:

1 *We consider the question whether making fuel from factory emissions is economically viable.*
2 *To look further, we learn the basic chemistry of such a process.*
3 *We construct an evolutionary model for efficiently selecting fuel-emitting microbes.*
4 *Next we turn to constructing a very basic (nonrealistic) profit–loss model for the business.*
5 *Short of answering the problem, we nevertheless gain an understanding of the chemistry and economics around it.*

Framing the Question

Reduce, reuse, recycle is a mantra of the green movement. Repurposing is another buzzword of the eco-friendly. We all see the value in *reducing* toxic emissions into the atmosphere, but could we go further and potentially somehow reduce, reuse, and repurpose the emissions themselves? This bold question is being pursued by an Evanston corporate neighbor, Lanzatech, located just down Oakton Avenue in Skokie, Illinois.

The basic idea is this: steel mills and oil refineries emit pollution in the form of gases. Among these gases is carbon monoxide (CO), which is a potential source of energy. (It's also deadly.) Would it be possible to make use of that energy?

This is a qualitative question. We should ask first if it is theoretically possible, and next if it can actually be done. To answer such a question in the affirmative, one could collect some of these gases and find a way to separate the carbon monoxide and convert it, somehow, into fuel. Even a small drop of fuel would lead to a yes response, a *proof of concept*.

But whether this is a *viable* source of fuel, and whether this could provide an ecological or economic benefit, depends on the details. What follow-up questions might probe the viability of this carbon capture?

Think of your response and jot it down.

By *viable*, we will interpret this to mean that a company that aims to repurpose pollution can succeed as a business. In other words, we have framed the issue by interpreting our driving question as follows:

> *Is it economically viable to capture emissions and turn them into marketable products?*

What would an approach to answering this question look like? We first would have to understand and describe the processes involved at a theoretical level. Then we need to gauge what is even *possible*, technologically speaking. Finally, we will want to try to price things out to determine whether the (economic) benefit is worth the cost. So we will study the question from scientific, technological, and economic perspectives. Be aware that this is a highly complicated issue and we are wading into deep waters – but that's a good way to get our feet wet! In the end, we will only address a small selection of simplified problems, hopefully gaining an appreciation for the work of many scientists, engineers, financiers, and entrepreneurs. We will not try to create a realistic business plan for a sizable company.

Let us emphasize: immediate economic gain is only one measure of feasibility, even if we just limit our perspective to economics. A country may decide that it is worthwhile to perform this process even if it comes at a short-term economic loss. Such a country might adopt the longer-term viewpoint that having a healthier planet, climate, and population not only is the right thing to do but will prevent further expenses from the ill effects of pollution. In short, there is economic benefit to our well-being. Climate change due to emissions can also be seen as a security and international relations issue, with new climate refugees and increasingly scarce resources driving geopolitical events – events that themselves have economic implications. But we will adopt the narrow view of an investor: might there be a return for an investment in such technology?[1]

Basic Chemistry

If you want to create a fuel for a combustion engine, you will need to know what happens to it when it burns. "Burning" actually means combining with oxygen, so the process of burning is a chemical reaction in which your substance is combined with oxygen to produce new compounds.

Let's recall that atoms are made up of a positively charged nucleus surrounded by negatively charged electrons in orbit. The nucleus itself is a massive dense core comprising positively charged protons and electrically neutral neutrons. This "Bohr model" of the atom resembles our solar system: a massive sun surrounded by orbiting planets attracted by gravity – only in the atom, it is the electrical attraction of oppositely charged particles that keeps the electrons in orbit. Atoms

---

[1] Disclosure: I know the chief scientific officer of Lanzatech, which is how I found out about it. The point of this chapter is *not* to convince readers to invest (!) but to demonstrate an interesting mixture of quantitative arguments from chemistry and economics, in action. I myself have no stake in the company.

are overall neutral, so there are as many electrons in orbit as there are protons in the core. This number is called the "atomic number," and the elements are listed in the *periodic table* in order of their atomic number (chemical symbols in bold): Hydrogen (1), **He**lium (2), **Li**thium (3), **Be**ryllium (4), **B**oron (5), **C**arbon (6), **N**itrogen (7), **O**xygen (8), **F**luorine (9), **Ne**on (10) ... Atoms can bond together to form *molecules*, which may be combinations of a single element – such as a molecule O_2 comprising two oxygen atoms – or two or more atoms, such as H_2O, a molecule of *water* comprising two hydrogen atoms and one oxygen atom. From these examples it should be clear what the subscripts represent. Note that the first letter of a chemical symbol is capitalized (or the only letter, if there is just one), so when you see an expression like CO, you can tell that it means carbon and oxygen rather than cobalt (Co). In fact, CO is carbon *mon*oxide, while CO_2 is carbon *di*oxide, as it carries two oxygens (the prefix *mon-* means "one," while *di-* means "two").

Basic to the subject of chemistry is the notion of a *reaction*, where the bonds within some molecules are broken, atoms are interchanged, and new compounds emerge. Some reactions require energy to proceed, and some release energy as a product. (Both can happen at the same time: if you toss a ball over the rail of a balcony, it takes some energy to get it over the rail, but more is gained as it falls to the ground floor and picks up speed.) If, overall, energy is released, the reaction is called *exothermic*; if a reaction absorbs energy from its surroundings (usually in the form of heat), it is called *endothermic*.

Consider an example. As we mentioned, "burning" means combining with oxygen. What happens if you burn carbon monoxide? The input gases are CO and O_2. The output of this process is known to be carbon dioxide, CO_2. By adding up the numbers of atoms involved, you see that we couldn't possibly have a reaction like $CO + O_2 \longrightarrow CO_2$, because on the left side, there are three oxygens (one from the CO and two from the O_2), while on the right side, there are two. However, if we jiggle the numbers a bit (see Appendix 3, Exercises 9 and 10), we find the reaction

$$2CO + O_2 \longrightarrow 2CO_2.$$

There are two carbon and four oxygen atoms on both sides of the reaction: the equation is *balanced*.

In steel plants and iron smelting factories, carbon monoxide, a deadly gas, is produced because there is not enough oxygen around to burn down "all the way" to carbon dioxide. So factories "flare off" the remaining CO, i.e., burn it. The resulting emissions, carbon dioxide, are much less deadly – recall we emit carbon dioxide every time we exhale – but in large quantities, this contributes to global warming.[2]

---

[2] The science behind the "greenhouse effect" is not complex. The Sun emits light in various frequencies (colors). The light we see from the Sun is the light that has made it through the Earth's atmosphere, and appears yellow. The Earth absorbs light from the Sun, and this process converts the light into heat. The Earth also, being a warm body, radiates light just as a burning piece of coal does. But the light that the Earth radiates is in different frequencies than the light which is absorbed, which is why the Earth does not look yellow from space (the colors that the Earth appears from outer space are due to reflected light from the Sun, rather than radiated light; the

Most fuels are hydrocarbons, made of molecules comprising mainly carbon and hydrogen atoms. Plastics are also made of hydrocarbons. So since hydrogen is abundant (e.g., in water), it is conceivable that one could create hydrocarbon fuels from reactions with inputs CO and water. The company Lanzatech has developed a bacterial process using microbes that can perform the necessary chemical reactions as part of their life cycle: they eat the CO and spit out the fuel.

The process is described as a bit like fermentation, or brewing, wherein yeast eats sugar and produces gas. This is how alcohol is made for human consumption but also how ethanol is made. What makes ethanol less viable as a fuel, however, is that its source is vegetative – *food* – which already has a high value. Returning to the capture of carbon monoxide, the procedure involves combining in water the microbes and the CO-containing gases, then letting the tiny bugs do their thing. Here are some products that can be produced from carbon monoxide and hydrogen, with the corresponding chemical equations:[3]

Product	Chemical reaction
Ethanol (C_2H_6O)	$2CO + 4H_2 \longrightarrow C_2H_6O + H_2O$
Acetone (C_3H_6O)	$3CO + 5H_2 \longrightarrow C_3H_6O + 2H_2O$
1-Propanol (C_3H_8O)	$3CO + 6H_2 \longrightarrow C_3H_8O + 2H_2O$

The viability of your commercial enterprise will rest on how efficiently you can carry out these reactions. If the cost of running the machines and pumping the tanks and paying the workers is more than the value of the final product, then you won't succeed as a business. Therefore, beyond just the chemical theory, one must master the *technology* needed to create an efficient process. Next we shall explore but one small component of that process.

An Evolutionary Model

Not all microbes, even proprietary ones, behave the same. For evident reasons, having microbes that multiply more quickly will be more beneficial to the carbon-conversion process.

Imagine a broth filled with two kinds of bacteria that multiply at different rates. What would happen if you let them be fruitful and multiply until the broth was

---

frequencies of the radiated light are not in the visible part of the "color" spectrum). So here's the thing: carbon dioxide absorbs more of the light in the frequencies that the Earth radiates than in the frequencies that come from the Sun. So the more carbon dioxide in the atmosphere, the less heat can escape the Earth. It becomes "trapped" in the atmosphere and warms things up.

Of course, this argument is only qualitative – but we are not studying the greenhouse effect in this chapter, we're studying carbon monoxide capturing. Also, a warning: greenhouses don't actually become warm due to the greenhouse effect! They become warm for the same reason the inside of a car heats up in the sun: because the hot air is trapped by the closed windows.

[3] These equations represent net ins and outs; the actual processes may be more complex.

thick as pea soup, then diluted it with water and repeated the process? We will show that this serves as a kind of artificial selection process, preferencing the fast-multiplying bacteria.

If both species have a constant birthrate, their populations P_1 and P_2 will both be exponential functions of time. As we show, this is what "constant birthrate" *means*. Birthrates are given in units such as number of births per 1,000 members of the population per year. So a birthrate of 6 people per thousand members of the population per year will mean that a community of 3,000 would expand by $6 \times 3 = 18$ members in a year. A population of P will grow by $\frac{6}{1000}P$ in a year, a number proportional to P.

Exponential functions are defined by this characteristic: they grow (or shrink) by a fixed factor in a fixed amount of time. (See Appendix 6.4.) And just as interest rates can be compounded continuously, population rates can depend on a continuous time parameter. A population with birthrate β will grow as $P(t) = P_0 e^{\beta t}$, where P_0 is the population at time zero (again, see Appendix 6.4). Here the birth rate is written as births per single member of the population per unit of time – so in our example, $\beta = \frac{6}{1000}$, not 6. Note, too, that we can assume that the birthrate is a "net" quantity, accounting for both births *and* deaths, the latter contributing negatively.

Therefore, if the first species with population P_1 has A initial members and a growth rate of β_1, and the second species with population P_2 initially has B members and expands with a constant growth rate β_2, then we know that

$$P_1(t) = Ae^{\beta_1 t}, \qquad P_2(t) = Be^{\beta_2 t}.$$

Let us decide at the outset to call the species with the greater birth rate the first species, so $\beta_1 > \beta_2$; and of course we take $A > 0$ and $B > 0$, or else the populations would be negative or zero, which is either nonsensical or uninteresting.

> *Given these data, write the fraction of bacteria represented by Species 1.*
>
> *Think of your response and jot it down.*

The question is simple if we ignore the time dependence for a moment. The total number of bacteria is $P_1 + P_2$, so the fraction which are the first species is $\frac{P_1}{P_1+P_2}$. Dividing numerator and denominator by P_1 allows us to express this as $\frac{1}{1+P_2/P_1}$. Recalling the actual functions allows us to write this as $\frac{1}{1+Be^{\beta_2 t}/Ae^{\beta_1 t}}$. Algebra and exponential rules give this as

$$\frac{1}{1 + \frac{B}{A}e^{(\beta_2-\beta_1)t}}.$$

But since $\beta_1 > \beta_2$, the quantity $\beta_2 - \beta_1$ is *negative*, and thus $e^{(\beta_2-\beta_1)t}$ represents a decaying exponential function. So after a long period of time, this term will approach zero. So the fraction of bacteria represented by Species 1 approaches the value $\frac{1}{1+0} = 1$ over time: the fast breeders take over.

In practice, we can't let the bacteria grow forever in the same broth, as they will run out of food – but since we are considering the *proportion* of bacteria represented by Species 1 over time, we can periodically dilute the broth without changing this fraction. As a result, this process weeds out the slower bacteria and leaves us with the fast-multiplying bugs that we desire.

Having fast-growing bacteria will speed up the "fermentation" process and allow the operation to be more efficient and productive. This will lead to a more profitable business model.

An entrepreneur will also explore other technological advances, such as making the process more tolerant to different concentrations in the source as well as the presence of possible contaminants in the brew.

To make these statements about efficiencies and profits more *quantitative*, we should construct a financial model of the business. In the following, we will not make a rigorous case study, as one might in business school or actually in business. Instead, we'll content ourselves with a simplified yet instructive model.

Economics

Economics is adding up the revenues and subtracting the costs. The bigger the difference, the more profit you make. You can therefore increase your profit in two ways (or a combination of the two): lower your costs or increase your revenues. For example, increasing efficiency with the selective breeding process discussed above will lower costs. To get an estimate of revenues, we make a careful financial model of income and expenses.

So, with our basic understanding of the chemistry – some technology involving a microbial process to convert carbon monoxide gas to hydrocarbons – let's try to see if we can make a business out of carbon capture. To address economic feasibility, the most direct method of analysis would be to list all the costs and potential sources of revenue, then compare. This might require some brainstorming.

Think of three possible expenses and write them down.

Stripping away all the science and fancy economic speak, in order to sell a bowl of ramen, you have to get the ingredients, make the soup, bring the soup to the market, and find a hungry customer with money. In so doing, you incur other costs: you have to get a place in the market, you need pots and knives and bowls, and you have to clean up after yourself. You might have to hire help. Some costs are ongoing and some are one-off expenses to get up and running. Before you even start, you need to get a pot and knives; you need to develop the perfect recipe; you need signage and a chef's hat. Some of these start-up costs recur: you might need to maintain/replace your cookware; you need to research new soups before your customers stray for the latest upstart.

Returning to our carbon-capture business, here is a table of possible costs.

Expense	Comment
raw materials	You need to procure the stuff that you'll convert to fuel.
transport	You need to bring stuff to your plant and get your product to the buyer.
operating costs	You need electricity, water, etc.; you need to pay workers and researchers
legal costs	There are scads of laws around energy production.

Here are some revenue streams.

Revenue source	Comment
sale of the product	You will make fuel, or some other materials, and sell it.
patents licensing	You may be able to charge other companies to use your process.

A quantitative answer to the question *Is it worth it?* would involve estimating the costs and revenues and seeing if you have a net positive. But to do so, you need to create a production model: will you be making one jar of jet fuel or enough to fill a lake? Will you be producing jet fuel exclusively, or a host of products? How much will all that cost, and how much do you realistically expect to earn? Even existing businesses will consider changing their production capacities and products, estimating profitability and feasibility in a number of scenarios. Circumstances are constantly changing, and businesses want to hit that moving target of maximum profit. This may require investment, and a bank or venture capitalist will need to be convinced through a sober analysis of the market.[4]

Building a Mathematical Model

For the purpose of instruction, in this chapter, we will not try to gain detailed figures but rather focus on how to build a model of costs. If we label the values in the tables above by the capitalized first letters R, T, O, L, S, P, then the profit, $\mathcal{P}$ (we have to use a new font, since "P" was already taken by "patents"), is revenues minus costs, or

$$\mathcal{P} = (S + P) - (R + T + O + L).$$

Now suppose our production model is to make two products in different amounts, say,

x units of jet fuel y units of electricity,

---

[4] A venture capitalists is someone who will give you money to help you start or expand your business, typically in exchange for partial ownership of the company.

in whatever form of measurement we choose: gallons, barrels, whatnot. More complicated models can be built, to be sure, but this one already has some salient features. For example, it may cost more in operating costs to make jet fuel, but you may make more in sales from it. But that may not always be the case. For example, if you have only one buyer for the jet fuel, you won't want to make more than the buyer is willing to purchase.

So each of the functions $S, P, \ldots$ depends on both x and y, and therefore the profit P is a function of x and y as well: so we want to figure out how to choose x and y so that $P(x,y)$ is maximized.

Typically (though not always – see "Scale" below), if x and y both double, then P will at least increase, if not double, so it would appear that we should make them both very, very large in order for P to be as large as possible. In practice, however, we can't just make tons and tons of jet fuel and electricity, due to the limitations of our facilities and our suppliers, or because at some point we may not find buyers for all that fuel. So we can't simply make x and y huge so that P grows. Instead, there is often an overall limit on $x + y$, or some such combination of them, so that $x + y \leq M$.

Let's make this more concrete. Suppose your facility can run for 24 hours. Suppose x is measured in barrels per day and y in megawatt hours per day.[5] Suppose it takes 4 hours to make a barrel of jet fuel and 1 hour to make a megawatt hour of electricity – and that your plant can only do one or the other at any given time.

What is the limit on x and y production?

Think of your response and jot it down.

The limit is imposed by the number of hours in a day. If we make x barrels of jet fuel per day and y megawatt hours of electricity per day, this will take $4x + y$ hours of operation, so $4x + y \leq 24$. And of course, $x \geq 0$ and $y \geq 0$. If we have determined by some other analysis that operating at capacity is possible and profitable in that we'll find enough buyers, then we'll want to max out and have $4x + y = 24$. We can easily solve this equation for y to determine

$$y = 24 - 4x.$$

Note this means x can't be too small: since $y \geq 0$, this implies $24 - 4x \geq 0$, or $x \leq 6$. That means that the profit P can be expressed as a function of x alone. We can then use graphical or other methods to determine where that function is maximal in the allowed range $0 \leq x \leq 6$.

It is easy to see how the finance of business quickly becomes complicated, even with relatively simple models where perfect information is available. In more realistic applications, there are many more variables, more uncertainty, and a constantly changing landscape of suppliers and customers.

---

[5] We picked units for these types of energies that typically cost within an order of magnitude of each other.

Let us further make the assumptions that your customer is also your supplier –
e.g., you will convert factory emissions into fuel for the selfsame factory – and
that you have no patents and no legal costs (risky, but that's just you). Since your
customer is giving you the raw material at the plant, there are also no transport
or material costs. In short, we assume $R = T = L = P = 0$. Of course, this is a
terribly simplistic assumption, but our point is to demonstrate the model. Although
a more realistic one will involve more components and more complexity, the basic
machinations will be the same.

With our basic model that we have only sales for revenue and operations for
expenses, we have $P = S - O$. Let us measure these units in dollars per (30-day)
month. Here O are the operating costs, and let's say, for example, that it costs $30
to produce a barrel of jet fuel and $5 per megawatt hour of electricity produced.
These numbers are meant to include all costs of operation, including labor, facilities,
energy, and everything else.

What is the function O then?

Think of your response and jot it down.

The answer is that the *daily* operating costs are $30x + 5y$, so the monthly costs are
30 times that:

$$O(x) = 30 \cdot (30x + 5y) \qquad \text{recall } y \text{ depends on } x$$
$$= 900x + 150y$$
$$= 900x + 250(24 - 4x) \qquad \text{since } y = 24 - 4x$$
$$= 900x + 150(24 - 4x)$$
$$= 300x + 3600.$$

Now let's model sales.[6] We suppose that the demand for electricity is constant,
allowing us to fix the price at $15 per megawatt. So if we make y megawatts per
day and sell each for $15, then the sales revenue from electricity is $15y$ in dollars
per day, or

$$30 \cdot 15y = 450y = 450(24 - 4x) = 10800 - 1800x$$

in dollars per 30-day month, where again we have used the fact that $y = 24 - 4x$.
Now let us suppose for the sake of argument that the demand for jet fuel declines as
we make more of it, meaning that the average price we can get per barrel declines
as x increases. We'll suppose the price we can get is $80 - x$ dollars per barrel. Then
the daily sales from jet fuel are $(80 - x)(x)$, with monthly sales of

$$30 \cdot (80 - x)(x) = 2400x - 30x^2.$$

Total sales comprise the sum of the sales of jet fuel and electricity:

$$S(x) = (10800 - 1800x) + (2400x - 30x^2) = 10800 + 600x - 30x^2.$$

---

[6] The prices are not very far off national averages at the time of writing.

So we can calculate the total profit to be

$$P(x) = S(x) - O(x)$$
$$= (10800 + 600x - 30x^2) - (300x + 3600)$$
$$= 7200 + 300x - 30x^2.$$

The graph of this function is a parabola opening downward (see Appendix 6.3). It has its maximum at $x = 300/(2 \cdot 30) = 5$. So profitability is greatest if we make $x = 5$ barrels of jet fuel, and therefore $y = 24 - 4x = 24 - 4 \cdot 5 = 4$ megawatts of electricity. The monthly profit is then

$$P(5) = \$7,950,$$

about \$8,000 per month.

We have created a business model and concluded on its basis that there was profit to be made. In some sense, then, we have answered the question we posed. But we note that in this model, the profit is quite small! We're certainly not going to become energy tycoons at this rate, nor will we put a dent in our nation's emission of carbon monoxide with these small rates of production. To make this a successful business and to successfully complete its green mission, we could hope to increase the capacity of the facility to produce fuel more quickly, or to repeat this process at many more facilities. Even still, at just about \$8,000 per month, we would have to spend a lot of time and energy (and start-up money) just to make another \$8,000 in profit. What makes the prospect of this viable is *economies of scale*, where our per-unit profitability can go up the more units we produce.

We therefore turn our attention now to understanding how processes can "scale."

Scale

"Scale" is an important concept in business. If you have a reliable way to make money, even if it is a very small amount of money, you can parlay that into big bucks if it "scales" well.

At first glance, you can just "double everything" (run your process twice, say) and you'd think you would make twice as much money, triple for triple the money, and so on. Of course, if you had to double all costs and the number of people working, the profit per person may be just as small. This is good, but only marginally satisfying: all that expansion will cost money, and it will be hard to attract investors with such modest gains.

But some processes scale well. If you run a hair salon and have the space in your shop to double the number of chairs, then, if demand is strong, you can consistently get twice the revenue (minus the start-up costs of installation) without paying a cent more in rent, and for far less than the double electricity/heating/janitorial costs. (The salaries you pay might have to grow if the hair stylists are your employees, or may not if you rent out the chairs and treat the stylists as independent contractors.) This means that your profits will rise faster as you grow the business.[7]

If you have a factory running one shift of eight hours each day, you might triple your output and sales by running three shifts, and yet even though you might have to triple the costs of materials, fixed costs like rent won't increase; maintenance may increase but likely not threefold; and factory worker salaries might triple (or a bit more if you have to pay extra for the night shift), but salaries to executives and most managers likely wouldn't.

Let's consider a concrete, albeit only slightly realistic, example. You own a company that makes medical devices for hospitals – you know, the machines that beep. You employ salespeople to visit hospitals and help install the beeping machines (part of their income is paid by you, part comes from a percentage commission on the sales they make). You employ support staff to assist the hospitals when necessary. Here is a breakdown of payments, in thousands of dollars per month:

- advertising: 5
- salesperson and staff salaries: 20
- assembly workers: 50
- materials (to build the machines): 20
- managers salaries: 10
- rent/mortgage: 20
- transportation/travel (mainly for the salespeople): 10
- electricity, maintenance: 5

For this model we suppose that you are able to sell 20 devices per month, each priced 10 (still in units of thousands of dollars), for monthly revenues of 200. Your monthly profit is then

$$200 - (5 + 20 + 50 + 20 + 10 + 20 + 10 + 5) = 60.$$

The profit per unit sold is $60/20 = 3$ thousand dollars.

What if you doubled production and doubled your sales force? Maybe this is what the new production model would look like:

- advertising: 8   (a wider range, but you don't have to make new ads)
- salesperson and staff salaries: 40   (this doubles)
- assembly workers: 80   (machines will work double, but not all of the workers)
- materials: 35   (you negotiated better prices for your larger order)

---

[7] Some businesses, such as jewelers, make a lot of money with each sale, but not so many sales. Some businesses, such as grocery stores, make very little money with each sale, but sales are robust and the multiplicative effect allows them to achieve good profits. This is called "making it up in volume," i.e., the volume of sales is large. The old joke is that the retail clothing outlet is being squeezed by the department store, so the owner decides to have a sale to generate business. They lower the prices so much that the manager complains, "But boss, at these prices we're losing money with each sale." The boss replies, "That's okay, we'll make it up in volume!"

- managers salaries: 14   (more, but there are no new processes to govern)
- rent/mortgage: 25   (a bit higher to pay for the small expansion of your factory)
- transportation/travel: 25   (farther range; more than double the cost)
- electricity, maintenance: 8   (heating, lighting would increase but not double)

As you expanded your territory, you found some hospitals with beeping machines in fine working order. To seal the deal and help grow the business to new customers, your had your sales force offer some bargains. As a result, you were able to double sales, but in the end, you lowered the average price of the machines to 9.5.

So in the end, you sold 40 devices per month (double), but for revenues of just $40 \times \$9.5 = 380$, as you had to cut a few deals to make sales to new clients. In total, your profit is now

$$380 - (8 + 40 + 80 + 35 + 14 + 25 + 25 + 8) = 145.$$

Your profit per unit is now $145/40 = 3.625$ thousand dollars.

Note that expanding your production allowed you to make higher profits per unit, even at lower prices. This is one example of the economy of scale! (We note: depending on the business, any increase in profit may be desirable as the business grows, even if per-unit profits decrease.)

Looking more closely, we see that automated processes scale better than human processes. It costs twice as much to double the working hours of a human sales force, for example, but the same does not apply to machines. You can see why e-commerce has exploded.

Note that the balance was delicate: if the increased transportation costs could not be accommodated in some way, the company would not have scaled well – so perhaps continuing to expand production and sales force might not allow this company to remain profitable indefinitely.[8]

We could imagine creating such an analysis for our hypothetical version of a carbon-capture company. While $7,950 per month may suit an individual operating a company out of a garage, as we have hinted above, entrepreneurs, investors, and environmentalists would want to take the idea (much) further.

Summary

We approached the question of recycling pollution from scientific, technological, and economic grounds. We learned the basic chemistry: carbon monoxide is a toxic gas produced in industrial processes, which can be converted through chemical reactions into fuel. We learned a bit about how chemicals are notated and how they change in reactions, including how to balance the number of atoms at both ends of the process. We then discussed the mechanism under consideration: microbial conversion of CO emissions into hydrocarbon fuels. That was basic science. To understand how this might be made feasible, we studied a process that artificially

---

[8] There are many ways to address this: you can build a second home office from which your sales force is dispatched, or you can allow remote salespeople to telecommute, for example.

selected the bacteria to preference the fast-breeding ones, hoping that this would make the process more efficient and, ultimately, profitable. To analyze feasibility on economic grounds, we then discussed how to build a cost–benefit analysis for our purported business, with various factors involved in production and sales. Finally, we concluded that real profitability as a company would depend on our ability to *scale* our model up, and we gained some insights into how a business may or may not be able to scale up its production profitably.

Surely, a serious effort to capture carbon entails far more analysis than we have attempted here, as well as far more science and economics. But our aims were modest: we got a taste of the issues and gained an appreciation for the many efforts of environmental entrepreneurs worldwide.

Exercises

1 Write a one-page summary of the carbon conversion process described in this chapter.

2 You may have heard that the burning of methane contributes to carbon dioxide in our atmosphere. The methane molecule has chemical symbol CH_4. *Burning* means combining with oxygen. The result of the process gives carbon dioxide and water.

   Write a chemical equation with all of these elements (inputs separated by a plus sign, then an arrow, then outputs separated by a plus sign). Now balance the equation so that the total numbers of each type of atom on each side match. (See Exercises 9 and 10 of Appendix 3.) This procedure is called *stoichiometry*.

3 2,3-Butanediol ($C_4H_{10}O_2$) is another chemical that is produced by the Lanzatech process. It is an intermediate product in the manufacture of such things as nylon and rubbers. The Lanzatech bacteria consume hydrogen gas and carbon monoxide and produce 2,3-butanediol and water. Write the chemical equation for this process, then balance it.

4 Hydrogen gas (H_2) and iodine (I_2) react to form hydrogen iodide (HI). The rate at which H_2 and I_2 turn into HI is proportional to the concentrations of both reactants. In this case, assume that there is an excess of H_2 present, so its concentration does not change significantly and can be treated as a constant.

   Write an equation for the change in the concentration of I_2 after a small amount of time. Specifically, call $[I_2]$ the concentration of I_2 and express the change $\Delta[I_2]$ in a small amount of time Δt, in terms of $[I_2]$ itself and a constant of proportionality k.

   Now write an expression for $[I_2]$ as a function of time, in terms of an initial concentration C and k. (**Hint:** see Remark 3.1 in Chapter 3. You will have an exponential function.) How long will it take for 90% of the I_2 to be used up? Express your answer in terms of k.

5 You are the chief executive officer (CEO) of a car wax company and are looking to sponsor a race car driver as a celebrity endorser. The appeal of celebrities is sometimes measured by a "Q score," a number between 0 and 100.[9] Your quants (financial analysts) have determined that among the group of drivers being considered, a Q score of Q will bring in an additional expected sales of $1 + Q/20$ million dollars each year. (So, for instance, each one will bring in at least another million in sales.)

  You give your vice president (VP) of marketing a directive to spend up to one month to scout various race car drivers (with various Q scores) and try to negotiate a deal with one of them. You must figure out an allowance to give to your VP, a maximum you will pay for the endorsement contract, and this will depend on what the driver's Q rating is. In order to ensure that the deal is not a loss for the company, you must choose this maximum to be the largest value so that the expected net profit from this venture is not less than zero. You may assume that profit is 40% of sales and that the yearly salary of your VP is $240,000. (You'll want your VP to strike a deal for less than the max, and you may hint that a bonus awaits if a good deal is struck.)

  What is the maximum allowance you should give to your VP to spend on the contract? (Remember: it will depend on Q.)

6 You are the CEO of a company that sells refrigerators. You earn revenue of $800 on each refrigerator sold. Start-up costs are $1 million, and you incur a fixed cost (due to materials, labor, electricity, etc.) of $100 per refrigerator manufactured. Because you have to expand distribution and transportation services as you sell more refrigerators, you also have a cost of $0.1x^2$, where x is the number of refrigerators sold. (We suppose that all this happens in a given year.) What is the minimum number of refrigerators your company needs to sell that year to become profitable? Is there a number beyond which your company becomes unprofitable again? What is your maximum profit?

7 Find the yearly operating expenses and revenues from the budget of a local municipality that you know. (In the US, you are entitled to view these documents as they are part of the public record.) Cite three actual sources of revenue and three expenses from this budget. (This exercise involves research.)

8 *Cap and trade* is a market-based approach toward controlling pollution. In this example, the government will limit ("cap") the total amount of carbon dioxide that a factory will emit at 100 tons per year. If your company emits less than that amount, you may be able to sell ("trade") the rights to emit the

---

[9] The Q score is obtained by querying subjects and measuring, of those who heard of a celebrity (or brand, label, etc.), the percentage who rate that celebrity as "one of my favorites." Other possible ratings were "very good," "good," "fair," or "poor," so note that the Q score is not an average: two celebs could have the same Q score even though one might be frequently rated "poor," as long as they have the same number of "one of my favorites" responses.

remainder. Or if you need to emit more, you would have to purchase the rights to do so from firms that have come in under the cap.

Suppose your company makes a profit of $9 million in a year while emitting 120 tons of carbon dioxide before cap and trade is instituted. Assuming the amount you pollute and your profit are proportional to one another, how much profit will you expect to make after your emissions are capped? Up to how much would you pay for the right to emit another 20 tons?

Now suppose that you decide that it's crazy to pay other people in order to pollute, so you decide to develop a more efficient production model. The new technology costs $3 million to develop and cuts pollution by 25%. How many years will it take to make the investment in clean production worth the while?

9 The Hat and Bag Shop produces two products, hats and bags. Each hat takes two hours to produce, and each bag takes one hour. (The store runs at all hours.) Hats cost $17 to produce and sell for $45. Bags cost $20 to produce, but since the market for bags is more crowded, they sell for $50 − b, where b is the number of bags sold per day. How many of each product should the store produce per day to maximize profit, and what is the maximum profit? (Assume that all hats and bags that are produced get sold.)

10 Run the numbers on a child's lemonade stand. Is it profitable? Will your conclusion affect your decision as a parent? (This exercise involves research.)

11 Referring to the evolutionary model from the chapter, determine how the proportion of bacteria represented by Species 1 changes over time when there are three species with birthrates $\beta_1 > \beta_2 > \beta_3$. What if there are n species?

12 What is arbitrage? Why can't we all just buy low and sell high? (This exercise involves research.)

13 Think of a topic for a project related to this question.

Projects

A Paper or plastic? Which is less "harmful" to the environment? Be clear about how you are measuring harm, as well as the assumptions and limitations of your model. Suppose you were driving to the store but forgot your reusable bag. You could either continue on your way, using four of the supermarket's plastic bags, or you could turn back to retrieve your bag, adding an extra N miles to your drive. Beyond what value of N would it make best environmental sense to continue without turning back?

B Methane is another gas that contributes to global warming through the greenhouse effect. In the production of natural gas, some methane is emitted into the atmosphere. At present levels, there is much more carbon dioxide in the atmosphere than methane. Yet it has been found that emitting methane has a comparatively *larger* effect than emitting carbon dioxide. Find out why, and construct an analysis of the relative *environmental* benefit or harm of natural gas production versus coal. The problem is complex; be sure to consider opposing arguments that some interested parties might make. Find an angle that you can explore quantitatively, to a certain degree.

C You want to open a coffee shop. Write a business model for this venture. Be sure to discuss rent, inventory, salaries, prices, locations, customers (what if they just sit and chat?), utilities, and any other factors that might be relevant. Use prices that are realistic today for Evanston, Illinois, or your local municipality.

8

Why Is It Dark at Night?

Appendix Skills: Algebra, Geometry, Units, Functions

OVERVIEW

So many stars shining so bright, so why is it ever dark? We review Olbers's assumptions and explanation of the paradox, then refer to some modern physics to understand where these assumptions fail.

1 *Olbers assumes that the stars are evenly distributed in an infinite universe. We review his argument why this leads to a bright sky.*
2 *We then learn some chemistry and physics relevant to this cosmological question and see which assumptions are inconsistent with what is now known science.*
3 *Next, we fix a direction of sight and study the expected distance we'd have to look out before seeing an obstructing star. We study a function we call* vacuity, *measuring how empty or starless the sky would remain if we only consider stars closer than some chosen distance.*
4 *Making an analogy to (negative) interest rates, we find the vacuity to be an exponentially decreasing function.*
5 *Studying this function, we see that the distance to the first obstruction is much larger than the size of the universe – so the sky is not nearly filled with stars and remains dark.*

Olbers's Paradox

Once upon a time, Carl Sagan popularized cosmology on the television show *Nova* by saying that there were "billions and billions" of stars in the universe. Neil deGrasse Tyson revived the show decades later, and the numbers were still just as large. As you probably know, the Sun is but one of those stars, and it lights up the sky. So with all of those stars out there, why is it dark at night?

What do you think?

Think of your response and jot it down.

This classic question is called Olbers's paradox.[1] The formulation of the question given by Olbers (1758–1840) is to assume an infinite, static, homogeneous universe.

---

[1] The name is a misattribution. Many scientists had considered the problem before Olbers (in particular, Digges in 1576) and made valuable contributions.

This means that the stars are fixed in the universe, that it is infinite in extent, and that there are on average as many stars here as there are there – or anywhere. Somewhat more formally, *homogeneous* means if you find some number of stars in a large region of fixed size at one part of the universe, then you will find about as many stars in a similar region at another part of the universe – like, if the universe were chocolate milk with space the milk and stars the powder, then the glass would be well stirred. So if one region is twice as large as another, it should have twice as many stars. That is, the density of stars is constant: the number of stars in a given region is proportional to its volume.

Olbers's argument was simple. He said that the night sky should be bright because each star is a finite size and therefore occupies a finite fraction of our night sky. If there were an infinite number of such, our sky must get filled by them and appear light.

> *Does this settle it?*
>
> *Think of your response and jot it down.*

The argument as stated requires closer inspection, since, for example, the stars could all be hiding behind each other in a row, such that the light from distant stars gets obstructed before reaching us. Such complete alignment seems very improbable, but even contemplating it has revealed that we should make another assumption: that the stars are strewn randomly across the universe.[2]

It *is* possible to have an infinite number of stars that do not align but together still wouldn't fill up the sky. One way this could happen is if there were, say, exactly one star closer than 1 light-year away, one star between 1 and 2 light-years away, one between 2 and 3 light-years, and so on. (A light-year is the *distance* that light travels in a year, about six trillion miles; it is not a unit of time.) This would produce an infinite number of stars, but the distant ones would appear smaller and together they would not fill the sky. It may be counterintuitive, but we are familiar with this kind of thing from elementary mathematics: the fraction one-third is written $0.333\ldots$ in decimal notation. But the expression $0.333\ldots$ literally means $0.3 + 0.03 + 0.003 + \ldots$, and so

$$\frac{3}{10} + \frac{3}{100} + \frac{3}{1000} + \cdots = \frac{1}{3}.$$

So an infinite sum of positive numbers may be finite. This arithmetic problem is not a precise calculation about stars, but it does suggest a way around Olbers's paradox: we might have an infinite number of stars, each occupying a finite fraction of the sky, that may nevertheless leave large sectors empty.

---

[2] While we won't explain precisely what random means, it should be intuitively clear that the chance of one distant star being obstructed by another is small (there is more unobstructed space than obstructed space), so the chance of two stars aligning is small, and the chance of *three* stars aligning is even smaller, and so on, so that the chance of a major alignment of stars is miniscule. In fact, for centuries, the alignment of stars has been a notion that signifies a very rare event.

We must note: this example violates the assumption of homogeneity. The more distant stars are in less dense regions of space, since there is a lot more volume in the region between 100 and 101 light years away than there is between 2 and 3 light years away, and yet both regions contain the same number of stars in this example. Put differently, a Ping-Pong ball made of plastic with some thickness has a lot less material than a beachball of the same thickness. *Under our assumptions –* constant density of stars, randomly scattered across an infinite universe – Olbers's argument that it should be bright at night could still be plausible.

But light from distant stars is less intense. Is it possible that the total intensity of light from stars at night – even under Olbers's assumptions – would be small enough so as to appear dark?

To investigate, we will calculate the brightness of the night sky due to stars that are about some fixed distance away from Earth, then consider stars twice as far, three times as far, and so on. We'll find that at each of these distances, the contribution to the brightness of the night sky is roughly *equal* – and, of course, nonzero. Adding *the same* nonzero number to itself infinitely many times does give an infinite number, as opposed to our previous example, where the numbers we were adding got smaller and smaller. So not only would the night sky not be dark but it should be infinitely bright!

Are we done?

Think of your response and jot it down.

We're not done, because we still haven't actually *made* the argument that stars at different distances contribute equally! We just *said* we would make such an argument. That said,

When we are done making the argument, what will we be able to conclude?

Think of your response and jot it down.

We will have built a mathematical model to answer a physical question. If the answer is not observed in reality, then the model itself must have been incorrect. So we must reject either the assumption of an infinite universe, or a homogeneous one, or both – or reject the way we are modeling the way stars shine.

In this short discussion, we have framed the question and made some strides toward building a model. Now let's make the argument we discussed.

An Olbers Argument

We begin with the assumption of an infinite, homogeneous universe, so we have a constant density of stars: the number of stars in a region of space is proportional to its volume.

Suppose we look out at stars 10 light-years away – well, not *exactly* 10 light-years, but say in some thick spherical shell, 10 light-years away. The rubber part of a tennis ball is a good example of a thick spherical shell. The volume of this shell will be roughly its surface area times its thickness.[3] If we do the same thing for a spherical shell 20 light-years away but still having the same thickness, then the ball will be about four times the volume – for when you double the radius, the surface area quadruples, but as the thickness hasn't changed, the volume will quadruple as well. In formulas, the surface area is the constant 4π times R^2, so at twice the radius, the area is the same constant times $(2R)^2$, and this is four times as large, since

$$(2R)^2 = 2R \times 2R = 4R^2.$$

Then a shell of the same thickness but at three times the radius (30 light-years) will have volume nine times as large, since $3^2 = 9$, and therefore contain nine times as many stars – and so on. But these stars are farther away, so the brightness each one contributes will be less. How much less?

If one star is three times as distant as another, how much less bright will its light be?

Think of your response and jot it down.

Figure 8.1 and the subsequent discussion below illustrate the answer.

The star emits light energy. We can capture all of it by surrounding the star with a sphere of sunbathing cats, or photovoltaic cells, or marigolds. The power that these (perfectly efficient) photovoltaic cells can recover will be the product of the intensity of the light times the area that the cells cover – in fact, *intensity* means

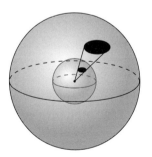

Figure 8.1 The large sphere is three times the distance from the center as the small sphere and has nine times the area. Corresponding areas, shown as black ovals, are illuminated by the same amount of light energy. The more distant region has nine times the area, and therefore the light that strikes it is one-ninth the intensity.

---

[3] The volume of a sphere of radius R is $\frac{4}{3}\pi R^3$. The volume of a sphere of radius $R + a$ is therefore $\frac{4}{3}\pi(R + a)^3 = \frac{4}{3}\pi(R^3 + 3R^2a + 3Ra^2 + a^3)$, which follows from expanding $(R + a)^3$ by using FOIL a couple of times (see Appendix 3). The volume of the spherical shell of thickness a in between radius R and $R + a$ is the *difference* of these quantities, $\frac{4}{3}\pi(3R^2a + 3Ra^2 + a^3)$. This value is of course *bigger* than its first term, $4\pi R^2a$, which represents the product of the surface area of the ball $(4\pi R^2)$ times its thickness a. However, if R is much larger than a, then the first term accounts for a very large fraction of the answer and so represents a good approximation. Anyway, if we just use this smaller first term in our calculations, then we will only *underestimate* the actual brightness from distant stars in our model. So if we conclude the brightness is infinite with an underestimation, then it will certainly be infinite using the greater quantity.

precisely power per unit area. From this we can tell that the intensity of the light shrinks as you get more distant.

Nonquantitatively, this is intuitively clear. If you want to charge up your glow-in-the-dark frisbee for a midnight game of ultimate, you hold it *closer* to your incandescent lightbulb. But we can also make a quantitative statement of the precise relationship between intensity and radius:

$$\text{intensity} \times \text{area} = \text{constant total power.}$$

Intensity is power *per* unit area. As we just discussed, the surface area of the sphere is proportional to the square of its radius, so the intensity will be *inversely* proportional to the square of the radius: $I = \text{constant}/R^2$. So to answer our question, if a star has some intensity at a distance R, then at distance $3R$, the intensity will be one-ninth as large.

So these stars that are about 30 light-years away – there are nine times as many of them, and the intensity of each is one-ninth compared to the stars 10 light-years away. These effects cancel perfectly. As a result, the overall contribution to brightness from the thick shell of radius 30 light-years is the *same* as the contribution from stars 10 light-years away. And the same argument works with 40 light-years away, 50, and so on, *ad infinitum*.

Now the brightness due to each of these shells is the same *nonzero* amount. The amount is nonzero because there *are*, in fact, stars. That is, the density of stars in the universe is constant (by our assumption) and nonzero by empirical observation (that is to say, stars exist). Therefore, each spherical shell of stars illuminates our sky equally, even if only slightly. So we have an equal, nonzero contribution from each of an infinite number of shells. The night sky should be infinitely bright!

This kind of conclusion could keep you up at night. But it doesn't – because it *is* dark at night, so we *can* sleep! Therefore, we must reject one of our assumptions: either the infinitude of space or the constancy of the density (or both), or find a flaw in our model.

Olbers was not privy to modern cosmology and the theory of a big bang, but his arguments led to suggestions of a finite universe. To understand some of the factors relevant to a modern resolution of Olbers's paradox, we'll want to understand a bit about how space, stars, and light work.

Warning: we will not perform any sophisticated analysis using the following synopsis, but we hope the reader finds it enjoyable and illuminating to read a crude distillation of hundreds of years of humanity's efforts to understand the natural world.

Cosmology, Physics, and Chemistry

- The space around you is three-dimensional, meaning if you put a ruler in one direction, then attach another at a right angle, then another at right angles to the other two, and so on, you'll only be able to do this for at most three rulers. This notion of dimension is equivalent to the fact that we represent a single number by a place in a number "line," that we use *two* coordinates to define a point in the *plane* and three for a point in space. Now Einstein realized that if we include time, we get a fourth dimension, and though this new direction is in some ways different, it is also wrapped up with the other three in an essential way. Warning: it is easy to visualize rotating the contraption of rulers within the three spatial dimensions, but involving the fourth dimension of time in this "rotation" is subtle. So after Einstein, we think of our universe as being a four-dimensional *spacetime*, blending the two words.

- If our universe were to comprise two spatial dimensions and one dimension of time, we might think of ourselves and our stars and galaxies as living on the two-dimensional surface of a balloon, evolving in time. Then we could imagine that some great overlord is blowing up the balloon, so that if you and your friends were parked somewhere, all the distances between yourselves would be increasing: the universe is expanding. This is a great visual that we can use for thinking about our own universe, which is *actually* expanding. It fails, however, because in our visual, the balloon was sitting in another space, whereas our four-dimensional spacetime is not evidently sitting inside some other (five-dimensional?) spacetime. No matter if our brains can't easily visualize it, expanding we are!

- The universe is large now but was smaller in the distant past, when "dust" started clumping together to form the stars and galaxies in the present universe. We live on Earth and feel the heat of the Sun, our nearest star and the core of our solar system. This solar system lives on the arm of a typical spiral-shaped galaxy comprising a few hundred billion stars, itself one of a few hundred billion galaxies.

> *Quick – about how many stars in the universe?*
>
> *Think of your response and jot it down.*

That makes at least $100{,}000{,}000{,}000 \times 100{,}000{,}000{,}000 = 10^{11} \times 10^{11} = 10^{22}$ stars!

- Stars in our universe emit *light*. "Light" is a word that includes the light you see and all microwaves, radio waves, X-rays, gamma rays, and any other kind of *electromagnetic radiation*. Just like pulses of sound are created by oscillating air and are measured by their pitch, or frequency of oscillation, light is measured by the frequency of oscillation of the electromagnetic fields (we won't go into explaining them here). The analogy between sound waves and light waves can be quite useful. When a race car passes by, you probably know the familiar *RREEEE-OOOOO* sound when it passes. The pitch of the sound goes down as it recedes away from you (the *OOOOO* part). The reason for this is that after the race car passes, the successive peaks of the sound wave take longer to reach you, so the frequency (the number of sound wave peaks reaching you each second) goes down. This is called the *Doppler shift*. A similar thing happens as light from a star travels toward you. The star is becoming more distant due to the expansion of the universe, so the frequency of the light which comes from it goes down. Since red light has a lower frequency than blue light, this is often called *redshifting*. This happens in all directions: since the universe is expanding, the stars are always receding, unlike the race car.[4]

- Let's talk a little about light and frequencies. The different colors have different frequencies. If you put a prism in a sunbeam, it turns out that the light at different frequencies will bend different amounts and get separated, producing a rainbow pattern. (In an actual rainbow, the droplets of water in the sky act as the prism.) If you put your prism in a pure beam of light coming from a neon or fluorescent light, however, you would not see the full spectrum of a rainbow but rather only certain distinct colors would emerge. Understanding why is how we are able to detect that the light that reaches us from distant stars has had its frequencies shifted. First we have to understand why stars emit light in the first place.

- You know those merry-go-rounds with the swings that fly out, or the carousels with the animals that you ride? The ones on the outside go faster, right? It's the same with the workings of an atom. An atom has a center, or *nucleus*, consisting of protons and neutrons, and the electrons orbit about it, the ones in the outer orbits flying around with more energy, just like the outer swings on the merry-go-round. If the atom is energized, more of the electrons will be in these outer swings. When the atom relaxes, the electrons move to an inner orbit of lesser energy. You may have heard that energy is conserved – so what happens to the excess energy, the difference between the higher-energy outer orbit and the lower-energy inner orbit? The answer is that the atom releases the excess energy in the form of light: a *photon* is emitted.

- I sneakily referred to light as both particle (by talking about a photon) and wave (by speaking of frequencies and peaks). Indeed, the dual nature of matter was a great mystery of the late 1800s, resolved by the theory of quantum mechanics. Part of this resolution is the discovery that each photon has a characteristic frequency, and its energy is proportional to the frequency: the higher the frequency, the higher the energy. (The constant of proportionality is called *Planck's constant*.) Another part of the resolution is that there are only a discrete number of swings available on the merry-go-round: not every continuous energy is accessible. That means in turn that there are only a discrete number of excess energies, as well. Toy model: if the possible energies are $1, 4, 9, 16, \ldots$, in some units, then the possible *differences* in energy are $3, 5, 7, 8, 12, 15, \ldots$. This means that the possible *frequencies* of emitted photons are discrete. We say that energy is *quantized*.

- Just as electrons release a photon of a particular frequency when they jump down an energy level, they can also absorb a photon of the correct frequency and jump

---

[4] Incidentally, the word *race car* is the same in both directions, so at least we have that!

up a level. Now, the list of possible differences in energy is unique to every atom. If a star, which emits light at a wide range of frequencies, contains lots of one type of atom in its outer layers, then these atoms will absorb photons precisely at this unique list of frequencies. When we analyze the frequencies of light coming from the star (for example, by passing the light through a prism), this shows up as a sequence of thin dark bands. By looking at the pattern of bands, we can determine the atoms in the star.

- Returning to our toy model, with its signature list of frequencies $3, 5, 7, 8, 12, 15, \ldots$, what if we actually observed $2.9, 4.9, 6.9, 7.9, 11.9, 14.9$? Unless there happened to be some other atom with this exact *new* signature, we could conclude that the frequencies had been shifted down by 0.1.

- If *all* the light from distant stars is redshifted as such, we'd conclude that all the stars were receding: the universe is expanding. On the flip side of this reasoning, in the past, the universe was smaller. Extrapolating, it would appear that at one point, the universe was tiny – and this intuition is borne out in calculations. The "start" of the universe is called the big bang and is well known in popular culture. Early in the universe, all the mass was in close proximity, and there was a virtual soup of photons. That light soup still permeates the universe, but the frequency and intensity has gone way down from all the expansion. The background of light soup glows at a temperature of about 3 degrees Kelvin, about -270 degrees Celsius. This is called the cosmic microwave background radiation (CMBR) and forms part of what "lights" up the sky.

- By analyzing the frequency signatures of sunlight, we know that our Sun contains hydrogen and helium. In fact, this is a hint toward a much more powerful mechanism behind the release of light energy: *fusion*. Hydrogen itself contains the raw ingredients for any atom: a proton and an electron (protons can smash together to produce neutrons – and *anti*-electrons!). You can imagine that some hydrogens would smash together to form some new combinations, and so on, eventually producing helium, with its nucleus of two protons and two neutrons. This is precisely what the Sun does, and light energy is released in the process, as we now describe.[5]

  Now here's the thing: if you add up the masses of the stuff going into such a reaction, the answer is *higher* than the masses of the stuff (helium) at the end. What happened to the missing mass? It was released as *energy*! You know that equation, $E = mc^2$? Well the c is just a constant there, with no substantive meaning. The gist of it is simply *energy is mass*. The missing mass is emitted as light energy, and its frequency depends on how much mass is missing. But the supply of hydrogen for this process is not infinite, and neither is the life-span of a star: eventually, it stops emitting light.

---

[5] Humankind has been able to create this reaction too, but nuclear fusion is not yet a viable source of energy: our machines that enable the process currently require more energy to run them than they create. Nuclear *fission* is a process where you reap the mass-energy by breaking apart a heavy nucleus. This is what occurs in nuclear reactors.

This is too much to absorb fully in just a few minutes. Furthermore, this abridged account is entirely qualitative: scientists who propose these physical laws must make *quantitative* predications about the behavior of atoms, and this is another magnitude of difficulty, as the devil lies in the details. Such disclaimers notwithstanding, this cherry-picked story of physics and chemistry tells us which features are most relevant for us to resolve Olbers's paradox:

1 Stars don't last forever.
2 The universe is finite.
3 The universe is expanding – so light reaching us is redshifted and lower in energy.
4 CMBR.

> *Which of these would* enhance, *rather than* diminish, *the brightness of the night sky?*
>
> *Think of your response and jot it down.*

Look Out!

While a complete, quantitative resolution of Olbers's paradox is not only outside the scope of this book but also, scientifically speaking, an unsettled matter, meaningful approximations can be computed incorporating some of the factors above. We shall perform one, due to Lord Kelvin.

Suppose we took the argument of Olbers and asked how far we would have to go in a straight line until we hit a star. How would we do that calculation?

> *Any ideas?*
>
> *Think of your response and jot it down.*

We couldn't possibly do it without any hard data, but which data would we look for?

To answer, let's take a page from the mathematician's book of tricks and try to answer a much simpler but related question. Suppose the universe was a *one-dimensional* straight line, with stars being points on that line. Then from your position in this universe, you could look only one way or another. If there is at least one star on either side of you, then your "sky" (consisting of only these two directions of sight) would be "filled" with stars. Here's a crude sketch of what this looks like, with you at the center:

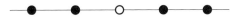

Since you can only look either right or left, there are just two stars in your visible sky: the closest ones to you on either side. What is the distance to the closest star? It's the same as the distance between stars. We could argue about whether you should be located in between stars or at the Sun in our one-dimensional world, but we'll see that the lesson can be learned either way. From this we see the crucial piece of data that we will need: we want to know how tightly these stars are packed in, i.e., their density.

We don't have realistic numbers to use since we don't live in one dimension, so we will use variables to represent the quantities we'll need. This will be most helpful for considering the three-dimensional case of interest. If there were ρ stars per unit distance, then we can figure out how large a region of our one-dimensonal space would be needed to contain approximately one star. Plugging in 1 for the number of stars and D for the size of the region, we see

$$\rho = 1/D.$$

Taking reciprocals of both sides, this says the distance D between stars will be $1/\rho$. With you located anywhere between stars, the average distance from you to the nearest star in your sky (the "lookout distance") will be $D/2$, or $1/(2\rho)$. This simple model already points to a feature of the three-dimensional setting: the lookout distance is proportional to the "volume" of space in which you would expect to find a single star, or inversely proportional to the density of stars (in one dimension, the "volume" is really just length). The feature that will be new in three dimensions is that stars have a *size*, a concept not evident in our one-dimensional model with point-like stars.

So much for the one-dimensional case. Let's up the ante a bit now. If the universe were a *two*-dimensional plane with you at the origin, then you could look out in a circle's worth of directions, measured by a compass direction. Each star in the universe would obstruct your view to some degree, by covering some fraction of those directions. How to calculate this fraction?

As you look out in two dimensons, you have a field of vision that is a circle. If a star of radius r is at a distance R from you, then it covers a segment of that circle. The size of the star is small compared to the interstellar distance, so r is much less than R. In this case, the star appears to be a line segment covering a length $2r$ (its diameter) along a circle of radius R. (Does that make sense to you? Talking about "line segments" as being parts of *circles* is a bit odd, but very large circles are approximately flat – just like the Earth, a large round thing, is approximately flat in comparison to sizes much smaller than its radius.) That is, each such star obstructs a fraction of the circle equal to its diameter ($2r$) divided by the circumference of the circle ($2\pi R$), or $\frac{2r}{2\pi R} = \frac{r}{\pi R}$. (See Figure 8.2.)

Precisely the same kind of question is relevant for the three-dimensional case: each star of radius r out at distance R shining in the sky will cover a fraction of

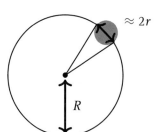
$\approx 2r$

R

Figure 8.2 In two dimensions, the fraction of the field of vision occupied by a star (gray circle) of radius r at a distance R from the viewer is approximately the diameter of the star ($2r$) divided by the circumference $2\pi R$ of a circle with radius R, or $\frac{2r}{2\pi R} = \frac{r}{\pi R}$.

it equal to $(\pi r^2)/(4\pi R^2)$, since the cross section of the star is a *circle* of radius r with area πr^2 and the total surface area of a sphere of radius R is $4\pi R^2$.

We now stay focussed on the three-dimensional case. There is a "quick and dirty" way to estimate the lookout distance. We'll do it first, then later complete a more detailed argument. First, note that the density of stars ρ, which already appeared in the one-dimensional calculation, measures the number of stars per unit volume. This means that $1/\rho$ measures the volume divided by the number of stars, or the volume per star. Let's put $V = 1/\rho$, the average volume of the "territory" occupied by one star. The units of V are length-cubed (e.g., cubic meters). The other quantity that seemed relevant above was the cross-sectional area, which has units length-squared (e.g., square meters). If we call σ this cross-sectional area of a star, then the lookout distance is expected to depend on V and σ and has units of length. The most natural combination of V and σ with these units is V/σ, or $1/(\rho\sigma)$. That's a crude *guess* at what the answer might be. In reality, there may be some coefficients, or other combinations of V and σ with the same units.[6] Still, this process of coming up with guesses based on units is very useful – so much so that it is given the rather aggrandizing term *dimensional analysis*. In our case, as we shall see from a more careful analysis, the naïve prediction is right on the money.

Stars and Cars

We ask, if we only look out to a total distance of R and no further, what fraction of the night sky would be vacant, unobstructed? By asking this question, we are creating a new *function* to encode this information which will be relevant to our example. The function – let's call it v for "vacuity" (or "vacancy") – will depend on an input variable, R, the radius out to which we are looking. So the function can be written as $v(R)$. Our aim is to determine this function $v(R)$, then see what its implications are for Olbers's argument.

Before doing any mathematics, we already know some important things about $v(R)$. When R is zero, there would be no stars yet to obstruct our view, so the full sky is unobstructed. This means $v(0) = 1$. We also know that as R gets large and approaches infinity, our view is cast to the whole infinite universe (we are still using Olbers's assumptions), and by Olbers's argument from earlier in the chapter, the whole sky is obstructed – so the unobstructed fraction becomes 0. We also know that v can only *decrease* as R becomes larger, since there will be more stars to obstruct our views, decreasing the vacuity. So $v(R)$ goes from 1 at 0 down to 0 at infinity. Yet still, we have no idea what the precise function $v(R)$ should be. To find out, we must make an actual calculation.

---

[6] For example, σ^2/V also has units of length, and we can cook up an infinite number of other such quantities. So the simple guess is only a guess. It will prove a good one, though.

In Chapter 3, we studied interest rates being compounded at different periods of time. We found that the loan balance changed in each such period of time by an amount proportional to the balance itself – the interest rate controlled the measure of proportionality. Over time, the total amount of the loan was determined to be an *exponential function*. In fact, in Remark 3.1, we found that if the interest rate is compounded continuously, the loan balance will be Pe^{rt}, where P is the initial amount of the loan (this is the case when you make no monthly payments back to the bank along the way).[7] Functions that *decrease* by a fixed fraction for each fixed time period exhibit exponential decay – see Appendix 6.4. The arguments are all precisely the same, except the "interest rate" r is negative.

We now argue why $v(R)$ is just such a function, namely, that if you increase R by just a bit, the vacuity decreases by a fraction of what it was at R. Enough prep: we are ready for the crux of the argument, establishing that the *change* in vacuity is proportional to the vacuity itself.

So let's suppose at distance R the vacuity is measured to be v, reprsenting the fraction of the sky that is empty or "dark," i.e., unobstructed by stars within a distance R. If we increase R a bit, say, an amount Δ, then what will the new value of v be? The value of v will *decrease* by the fraction of the sky covered by the stars that were *newly* revealed. Let's say that again. When R increases a bit, more stars will appear in the previously empty part of the sky, decreasing the vacuity by whatever fraction they cover. That fraction covered by new stars is given by the total area A that is newly covered divided by the total area. At distance R, that means $A/(4\pi R^2)$, since $4\pi R^2$ is the surface area of a sphere of radius R. The key will be to find A.

It's getting too abstract, so let's suppose $v = 3/4$ and there are 240 stars between radius R and radius $R + \Delta$.

> *How many stars contribute to new obstructions?*
>
> *Think of your response and jot it down.*

The answer is $240v = 180$, since only the fraction v of them in the vacant part of space will be new obstructions. From this we see that if there are S stars between R and $R + \Delta$, then there are Sv newly obstructing stars. If each star has a cross-sectional area σ, then the total new obstructing area is σ times this number, or $A = \sigma Sv$. So v decreases by an amount $\frac{S\sigma}{4\pi R^2}v$, and the fact that this is proportional to v already tells us that v is an exponentially decreasing function! (See Appendix 6.4.)

---

[7] Remark 3.1 explained that for a loan amount P with interest rate r, if you compound the interest n times in a year, you get a balance of $P(1 + r/n)^{nt}$ after t years. For continuously compounded interest, we let n approach infinity. The statement that the balance is Pe^{rt} then follows from an important identity that requires the calculus of limits to prove: as n approaches infinity, the quantity $(1 + r/n)^n$ approaches e^r.

Even better, the number of stars in between R and $R + \Delta$ is the density ρ times the volume of this region, which is $4\pi R^2 \Delta$, the surface area times the thickness. So $S = 4\pi \rho R^2 \Delta$, and the surface areas cancel to give that v changes by $-\rho \sigma v \Delta$.

To compare with interest rates (see again Remark 3.1), think of v as a loan balance and Δ as a small fraction of a year: an annual interest rate of $-\rho \sigma$ would mean a change in loan balance of $-\rho \sigma v \Delta$ in time Δ. In other words, v changes by a fractional rate $-\rho \sigma$. We can conclude, by strict analogy with the examples of interest rate and exponential decay, that $v(R) = e^{-\sigma \rho R}$.

In fact, the argument only allows us to conclude that v is proportional to this quantity, but since we know that when $R = 0$, we have $v = 1$, the formula is correct as written.

Good job! Time to put our feet up? Well, yes and no. We did find the function that contains the information we need, but we still need to harvest that information. Finding the function was a stepping-stone in the solution of the problem. Specifically, we have to figure out how far we have to go before hitting a star. Spoiler alert: we'll find that that distance is way larger than the universe itself!

To find how far we can look, consider some other possible functions that v *might have* been. If v were 1 (completely unobstructed) for all distances below 10 light-years, say, then equal to 0 at 10 ly, what would that mean?

Think of your response and jot it down.

It means that the spherical shell at 10 light-years is completely filled by stars, and so the lookout distance would be precisely 10 light-years from our perspective. (This example is not consistent with a homogenous universe.) You can understand, then, that if v goes down from 1 to 0 very quickly, then the lookout distance is small, while if v goes from 1 to 0 slowly, the lookout distance is large. So the lookout distance is a measure of how quickly v goes from 1 to 0. For the exponential function $e^{-\rho \sigma R}$, the lookout distance is a measure of its width: how long until the function decreases appreciably. As we can learn from Figure A6, for an exponential function, the width is easily determined: for us it is $1/(\rho \sigma)$, precisely our crude guess by dimensional analysis![8]

Summarizing, we have computed that the lookout distance in three dimensions is

$$\text{lookout distance} = \frac{1}{\rho \sigma},$$

where ρ is the density of stars in the universe and σ is the cross-sectional area of a star.

---

[8] An exact calculation is possible, though it requires some calculus. If $v(R)$ is the vacuity, then the fraction of the sky from newly obstructing stars between R and $R + a$ is measured by $v(R)$ minus $v(R + a)$. This can be thought of as the probability that when you look out, your view will be obstructed by stars in this range. Considering all ranges, the expected obstructing distance can be calculated just as in Appendix 7.1. The result is exactly $1/(\rho \sigma)$.

So what is it? To find out, we need data. There are about a hundred billion (1×10^{11}) stars in a galaxy and about a hundred billion (1×10^{11}) observable galaxies. The data are quite rough (estimates based on observations of the Hubble telescope), so we only carry one significant digit – see Appendix 5.3. Another rough assumption is that the Sun is a typical star. It has radius $R_\odot = 7 \times 10^8$ m. The radius of the observable universe is about 50 billion light-years.[9] The volume is roughly the cube of this number (again, we don't need to be precise, for reasons that will become clear), or about 1×10^{32} ly. We get a density by dividing:

$$\rho = \frac{\text{number of stars}}{\text{volume of universe}} = \frac{10^{22}\text{stars}}{10^{32}\text{ly}^3} = 10^{-10}\text{stars per cubic light-year.}$$

Plugging in the Sun's radius $R_\odot$ gives a cross-sectional area of $\sigma = \pi R_\odot^2 \approx 2 \times 10^{18}$ m^2. Now since 1 ly $\approx 10^{16}$ m, we have 1 m $\approx 10^{-16}$ ly and $\sigma \approx 2 \times 10^{18} \times 10^{-32}$ ly$^2 = 2 \times 10^{-14}$ ly^2. Plugging in, we find

$$\lambda = \frac{1}{\rho\sigma} = \frac{1}{10^{-10} \times 2 \times 10^{-14} \text{ ly}} \approx 5 \times 10^{23} \text{ ly.}$$

This is our lookout distance. But just above we cited the radius of the observable universe to be 5×10^{10}ly.

How much larger than the radius of the universe is the lookout distance?

Think of your response and jot it down.

Since $\frac{5 \times 10^{23}\text{ly}}{5 \times 10^{10}\text{ly}} = 10^{13} = 10,000,000,000,000$, we see the lookout distance is ten *trillion* times larger than the radius of the universe! In other words, we expect easily to look to the edge of our (finite) universe and see no star.

Thus, the finite size of the universe gives one resolution of Olbers's paradox: when you look out at the entire universe in some direction, you are very unlikely to see a star.

Other Arguments

Other arguments help explain why the brightness cannot be infinite, as well.[10] Kelvin recognized that there is a finite amount of energy in a star, and therefore it cannot radiate light indefinitely. If the would-be obstructing star had ceased to emit light, it would not have contributed to the sky's luminosity.

---

[9] You may ask how the universe can be 50 billion light-years = 5×10^{10} ly in radius, yet only 14 billion years old, without violating traveling the speed of light. The answer comes from the fact that the universe itself is expanding.

[10] The "argument" that the universe is finite *because* we have a dark sky is putting the cart before the horse. Saying that cosmology must have been created to be consistent with life on Earth – even though that life came way afterward – is known as the *anthropic principle*. Despite its fancy name, it is a truism to argue that life as we know it is compatible with reality.

A lesser factor is redshifting: since light from lower frequencies has less energy, and light from stars is redshifted (with light from distant stars being redshifted more), calculations of light intensity from distant stars must be adjusted.

For a lengthy discussion of this problem, see Harrison's book,[11] the source for some of the material in this chapter.

Conclusions

So, what did we learn? We first understood the paradox by computing the intensity of light hitting Earth from stars at a given distance. Assuming a constant density of stars, we found that the answer was some finite amount that did not depend on the distance away – so an infinite universe would lead to an infinite total intensity and a bright night. This was a first hint at a finite universe.

To approach the question anew, and understand why it's dark at night from a quantitative perspective, we had to estimate how much light from space strikes the Earth.

Before trying to resolve the paradox, we took a crash course in physics to understand the mechanisms by which light is created in a star and travels across a dynamic spacetime – then we sorted out the most relevant factors. (Part of the fun of addressing these questions is playing journalist for a day, diving headlong into unfamiliar disciplines.) We settled upon the finite size and age of the universe, the limited lifetime of stars, the redshifting of light, and the cosmic microwave background radiation.

While unable to deal with all of these factors, we were able to make a potent calculation based on a finite universe. We computed that under Olbers's assumptions, the stars occluding our view would have to have come from way beyond the edge of the known universe. The time the light would have traveled to do so is far greater than the age of the universe. So the obstructing stars are not actually there to shine, and our night sky is dark. The finite scope of the universe resolves Olbers's paradox.

Exercises

1 A *light–year* is the distance that light travels in a year. The speed of light is 3.0×10^8 m/s. What is a light-year in meters? In miles?

2 In astronomy, the unit AU for "astronomical unit" represents the distance from the Sun to the Earth, about 93 million miles or 150 million kilometers. This is a useful unit for measurements within our solar system, but beyond our solar system is the rest of the Milky Way galaxy, which is much larger, and other galaxies and galactic clusters that form the universe. To speak of distances on

---

[11] Edward Harrison, *Cosmology: The Science of the Universe*, Cambridge University Press, 2000.

such vast scales, we employ the unit parsec,[12] which is the distance at which a length of one AU subtends one second of arc, i.e., spans 1/3600 degree. Draw an isosceles triangle that demonstrates this definition. Use your picture to calculate what one parsec is as follows. First recall that 360 degrees is the same as 2π radians. (So one radian is $360/(2\pi)$ degrees; one degree is $2\pi/360$ radians.) Now the "small angle approximation" allows you to say, for a very long isosceles triangle such as this one, that the number of *radians* of the small angle is equal to the ratio of the short side to the long side. Use this to figure out what a parsec is in AUs and, finally, in terms of meters. Do the same for a megaparsec.

3 The radius of the Sun is about 7.0×10^5 km. The distance to the Sun is about 1.5×10^8 km. What fraction of the sky is covered by the Sun?

4 We often read about average temperatures. This question explores what that means. Suppose our "world" consists of a perfect two-dimensional square with sides of length 10 miles. The world weather service has been allocated enough money to fund the establishment of 100 sensors that can be programmed to measure temperature at a chosen time. Where would you put them to get a sensible measure of "average temperature"? Be specific by using coordinates to indicate the points where you would place them. (**Hint:** start with a picture.) With the sensors in place and scheduled to take a simultaneous reading of today's temperature at noon, how would you calculate the average temperature from the data collected? What potential sources of error would there be in your model? Can you think of two scenarios that "should" have markedly different average temperatures but would be treated as equivalent in your analysis?

5 If I fold a piece of paper once, I double the thickness. If I fold it twice, it's four times as thick as when I started. Imagine a really big piece of paper that you could fold as many times as you wanted. About how many folds would it take until the thickness reached from the Earth to the Moon? (You'll need to estimate the thickness of a piece of paper, somehow. Try to do so without research.)

6 Earth is the only known inhabitable planet, but which other planet would you find the most hospitable – or least inhospitable? Give reasons why. (This exercise involves research.)

7 Einstein's theory of general relativity is one of the most profound and subtle constructions of physics. Yet it has a practical application that many of us

---

[12] The name is a blend of "parallax second." *Parallax* means the angular distance as viewed by an observer, while *second* is a specific angle: a "minute" is 1/60 of a degree, and a "second" is 1/60 of a minute.

enjoy each day. Find out what I'm talking about, and write a few words about it. (This exercise involves research.)

8 Think of a topic for a project related to this question.

Projects

A Could we solve the world's energy and environmental problems by covering the Sahara with solar panels? Do a detailed analysis of the amount of energy available as well as the start-up and logistical costs. Include as well a discussion of relevant geopolitical concerns.

B The Nobel Prize–winning physicist Wolfgang Pauli in 1930 postulated the existence of an electrically neutral elementary particle called a *neutrino* (later found in 1956). Basically, because it does not react to electrical charge, it is only felt when it hits another particle head-on. For this reason, Pauli lamented, "I have done a terrible thing. I have postulated a particle that cannot be detected." Discuss the difficulties in detecting neutrinos, including especially a thorough accounting of their *mean free path*.

9

Where Do the Stars Go in the Daytime?

Appendix Skills: Functions

OVERVIEW

We approach this question in our usual manner: we familiarize ourselves with the terrain so that we may identify a relevant point that we can address quantitatively – in this case, how the eye's sensitivity is affected by the level of background light intensity.

1 *We first recognize that this question is about biology, not astronomy.*
2 *Next we familiarize ourselves with the workings of the human eye, particularly how it can react to changes of intensity across many orders of magnitude.*
3 *We then try to model how the eye might respond to light of varying intensity, concluding that the response cannot be linear.*
4 *We find that a logarithmic model makes good sense near some background and also explains why certain items that are viewable in one setting are not viewable in another.*
5 *We use data from the human eye to model this. Our model is sound: it shows how, to no one's surprise, stars do seem to disappear in the daytime and return again at night!*

Biology and the Eye

The night sky is filled with stars, but they're not there during the day. Does that mean that during the daytime, we happen to point to a starless region of the universe?

> *Think of your response and jot it down.*

In fact, the stars are still there, but we can't see the light from them due to the small *contrast* of daylight-plus-star relative to simple daylight. In other words, although the question seems to be about astronomy, the effect is actually *biological.*

In this chapter, we will try to develop enough understanding of vision to address this question from perspectives both qualitative (Why does the star disappear?) *and* quantitative (How bright would it have to be to be visible?).

Interpreting the problem in this way, we have already identified some areas to study. So after learning a bit about how the eye works and responds to stimuli in different settings, we will be able to address the question of the vanishing star.

So let's take a crash course in vision, starting from the basics.[1]

- As animals, we perceive the outside world and respond to it with thought or action. Thought will play no role in whether we see a star, so we can ignore questions of cognition and concentrate on perception.
- We perceive with sensation. We have five major senses: seeing, hearing, tasting, smelling, and touching. Each sense is governed by parts of the body that receive information from the outside world. We will focus on vision, meaning we study the eye. As we do, we may imagine studying parallel mechanisms that govern the other senses: the details would be different, but common themes would arise.
- The eye is responsible for vision. It is a round orb with a *lens* in the front. The lens focuses and projects light rays that strike the front of the eye onto the *retina* in the back. The retina is a sensor. When it perceives light, it signals this to the brain through the nervous system (nerves), the byways for such signals. At the front line of photoreception are the *rods* and *cones*, two kinds of cells that have different areas of specialty.
- Rods are good for light sensitivity, while cones are good for acuity.
- These photoreceptors (rods and cones) initiate a signal that in its essence communicates "light is here" to the brain. We will want to know something about how that signal is transmitted.
- You may have heard that nerves transmit electrical signals. This statement is a bit misleading: the mechanism is nothing like electricity propagating through a wire, which travels at millions of meters per second. Instead, nerve signals travel at 20 m/s, still plenty fast over the scale of the human body. The signals travel when charged particles (this is the "electrical" aspect) move due to *diffusion*. Diffusion means going from where there are more to where there are fewer. In the kitchen, when you smell pizza coming from the oven, it is because the pizza particles move from where there are more of them (near the oven) to where there are few of them (near your nose). On the farm, if you open up a gate, cattle that are cramped in a paddock will meander through the gate to the empty field. In the nerve, the charged molecules (*ions*) travel through an *ion channel*. This gate appears at the cell membrane, and the farmer who allows just a few cows to shift fields is the cell membrane, which erects an electrical "keep out" sign by means of a voltage differential. We won't concern ourselves with the details of electrostatics here. Suffice it to say that the aggregation of charges on the nerve cell membrane can act as a barrier to the passage of further charged ions.
- Along the route of neural transmission, there are several channels through which ions will pass. The gates themselves can be triggered to open either by a chemical or voltage change. To avoid transmission loss from a dampened signal, the nerve is insulated (by "myelin"). At locations where there is no myelin, the signal is boosted by voltage-controlled gates called "nodes of Ranvier." Preventing transmission loss is crucial for a well-functioning nervous system. Multiple sclerosis

---

[1] Main source: D. Purves et al., *Neuroscience*, 4th ed., Sinauer, 2008.

is a disease in which the body's immune system attacks myelin, and the resulting scar tissue distorts or disrupts the signal, causing sensory and motor issues, among others. In fact, since the eye is so sensitive an organ, visual disturbances are one of the key early indicators of multiple sclerosis.

- So we understand a bit of the mechanism by which signals are transmitted through neurons, but how do they encode shape, color, and intensity? Shape is easy: the photoreceptors are essentially the pixels on the screen that is our retina. Where they respond is where the image is projected. (The brain re-creates a sensible understanding from the pixel data.) Color is encoded by the "cone" receptors, for the most part. There are three types of cones, sensitive to different frequencies of light. (This is analogous to the fact that color pixels on many displays are made up of red, green, and blue subpixels.) Together, their firing encodes the color of the light that strikes the region of the eye near these three cones. Intensity is measured by how frequently the photoreceptor fires off a nerve signal by creating an *action potential*. The firing rate can range from 0 to about 300 times per second. If you think of a signal as being like a single tap on the shoulder, then repeated, rapid tapping might indicate more urgency – or here, intensity.

So light hits rods and cones in the retina of the eye, which react by initiating nerve signals that travel through electrically controlled gateways along a nerve up to a brain. These nerves form the pixels of our retinal screen. The color of light is sensed by the three types of cones, fine structure is sensed by the rods, and the intensity of the light is measured by the firing rate.

Linear and Logarithmic Models

Your eye can respond to light within an enormous range of intensity, or *luminance*. Luminance is typically measured in the quaint units of *candles* per square meter. A candle or candela is a unit of power, fittingly enough about equal to the power from one candle. In these units, the range of intensity over which your eye can function goes from about 10^{-6} (nighttime) through 10^{-2} (moonlight) to 10^2 (indoor lighting) all the way to 10^6 (bright sunlight).

How much more intense is the light at the top end of the range versus the bottom?

Think of your response and jot it down.

To answer the question, we need to figure out which notion of *more* is most relevant. We could measure the difference, but the difference in values does not convey as much meaning as the ratio. We might say that an adult tree is five times as tall as a sapling, and this way of comparison paints a decent picture even if we don't know the precise heights involved. In the present case, we want to know *by what factor* the top level of intensity is higher than the bottom. Dividing top by bottom, we get $10^6 \div 10^{-6} = 10^{12}$ – so 12 orders of magnitude. The light at the top range is a *trillion* times as intense!

To appreciate this range of sensitivity to intensity, let us compare with the range of *frequencies* in the visible spectrum, that is, the different colors. Frequencies are measured in units of hertz, or wave oscillations per second. The reddest light we see is about 400 terahertz, or 400 trillion hertz, whereas the most violet visible light is about 800 terahertz. There is thus a difference of a factor of only 2 (not two orders of magnitude even, just plain old 2) between the high and low ends of visible frequencies. This is perhaps not surprising, as acuity to color is not as vital to survival – many of us, and many animals, thrive while color-blind.

As I write this, I am looking out the window on a stormy morning. A maple tree's branches are swaying in the wind. The leaves on the southern branches have a lighter hue than those shaded by the tree's crown. I see this out the window of my unlit room, where in the foreground the contours of my couch cushion reveal its embroidered floral pattern. All of this happens in dim light, probably about 10^{-1} candles. The different gradations I am describing occur within a range of about two orders of magnitude (a factor of 100), and my eyes easily pick up the differences in intensity in this range. Below this range, features are too dim to see; above the range, they appear uniformly bright.

Now recall that the eye records these variations of intensity through the voltage spikes sent out by its photoreceptors, and they can fire at rates between 0 and about 300 times per second. In other words, these sensors have a "discharge rate," and the greater the light intensity, the higher the discharge rate. This rate thus defines a *function* that depends on the light intensity, increasing when the intensity increases.

The simplest such function would be a linear function going from 0 to 300 over the range of visible light intensities from 10^{-6} to 10^6 candles per square meter, so let us suppose for the moment that this is the case.

> Can you write a linear function that goes from 0 at 10^{-6} to 300 at 10^6?
>
> Think of your response and jot it down.

To answer the question (see Appendix 6.2 for a discussion of linear functions), we compute the slope to be $\frac{300-0}{10^6-10^{-6}}$, which is 3×10^{-4}, to one significant digit (see Appendix 5.3). Let's call the discharge rate R, measured in nerve firings per second, and the intensity I, measured in candles per square meter. We want to write R as a linear function of the variable I with slope 3×10^{-4} as calculated above. We can use the point-slope form of the equation of a line, where here the known point has $I = 10^{-6}$ and $R = 0$. We get $R - 0 = 3 \times 10^{-4}(I - 10^{-6})$, which to one significant digit says

$$R \approx 3 \times 10^{-4}I$$

whenever $I \geq 10^{-5}$, allowing us to drop the 10^{-6} term.

But now we must admit that the response rate is *not* a continuous parameter. It is a whole number that can be $0, 1, 2, \ldots, 300$. So this function can't literally be correct. Strictly speaking, a continuous response rate R with slope 3×10^{-4}

increases by 0.0003 firings for every candle increase in intensity. We can write this as $0.0003 = \frac{\Delta R}{\Delta I}$. But a value of 0.0003 nerve firings does not make sense. The smallest nonzero increase we can sensibly speak of is 1.

> *A change of how many candles would generate one more firing?*
>
> *Think of your response and jot it down.*

For $\Delta R = 1$ and ΔI unknown, we solve $0.0003 = 1/\Delta I$, so the firing rate increases by 1 for each additional $1/0.0003 \approx 3300$ candles of intensity (to two significant digits). Then the graph of the firing rate would not actually look like a continually sloped ramp but rather like a staircase: 0 for the first 3,300 units, then 1 for the next 3,300, up to 3 at about 10,000, and so on.

> *Can you see why that would be absurd?*
>
> *Think of your response and jot it down.*

As we mentioned, the light intensity of the morning scene I witnessed ranged over two orders of magnitude from a baseline of about 10^{-1} candles – so in a range from about $10^{-2} = 0.01$ to $10^0 = 1$ candles. But if the eye only increases its rate of firing by 1 every 3,300 candles, starting at a rate of 0 at 10^{-6} candles, then there would be no detection of the minor variations in intensity of light from the underside of the leaf, from the contours of my cushion, from the shady part of the tree. In fact, since the upper intensity of that scene I described was just 1 candle, much smaller than 3,300, my photoreceptors never would have even reacted: the scene would look black!

In Appendix 6.5 we explain how logarithmic functions can describe variations over great ranges. Maybe the response of discharge rate to light intensity is logarithmic, accommodating such a large range of inputs. So our function that goes from 0 to 300 over 12 orders of magnitude may be logarithmic. We'll see that this is a better model, but we'll still run into problems. Consider if $R(I)$ were logarithmic, up to the addition of a constant – so $R(I) = a + b\log(I)$.

Similar to how we constructed our linear model, we can use the fact that we know the values of R at two separate points to determine the unknown constants a and b in our logarithmic model. If we put $R(10^{-6}) = 0$, this says

$$0 = R(10^{-6}) = a + b\log(10^{-6}) = a - 6b,$$

where we have used the property $\log(10^x) = x$. Now if we set $R(10^6) = 300$, then we find

$$300 = a + 6b.$$

The first equation says that $a = 6b$. Using this information in the second says that $300 = 6b + 6b = 12b$, which gives $b = 25$ and $a = 6b = 150$. Therefore,

$$R(I) = 150 + 25\log(I).$$

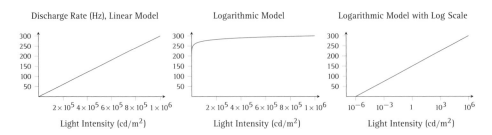

Figure 9.1 Possible discharge rate functions (nerve firings per second, or hertz): linear (left), logarithmic (center), and logarithmic with log scale in the x direction (right).

Could this be right? Let us ask,

> *How much would this function vary over two orders of magnitude of intensity?*
>
> *Think of your response and jot it down.*

To answer the question, note that over two orders of magnitude, $\log(I)$ changes by 2. Indeed, a range of two orders of magnitude goes from some number x to the value $100x$, and we use log rules (see Appendix 2.5) to compute $\log(10^2 x) = \log(10^2) + \log(x) = \log(x) + 2$: it increased by 2. So $R(I)$ increases by the amount $25 \cdot 2 = 50$.

All of this means that within a light range of two orders of magnitude, as in the scene I described, our neurons would only range over 50 of their 300 possible response rates. That is a *possible* solution to the problem, but one that loses sensitivity by exploiting only a sixth of the full range of neural responses. (The linear and logarithmic functions discussed so far are shown in Figure 9.1.)

Light Adaptation: Setting the Baseline

These functions are therefore not optimal, and in fact Nature has rejected them in favor of a far more clever solution. The eye actually performs a kind of magic trick known as *light adaptation*. When you walk out from a root cellar into bright sunlight, you are blinded while your eyes adapt to the change of background intensity: until they do so, your rods and cones are firing at the maximum rate. When you walk back inside, it's hard to see the dust bunnies in the shadows because hardly any of your sensors fire until the baseline is reset back down.

In fact, after the adjustment, the response rate is approximately logarithmic – but the actual logarithmic function *changes* according to the background conditions; i.e., there is no single function $R(I)$. What happens is that the muscles in the eye "jitter" so that the receptors sample the luminance not only of a spot but of surrounding regions as well. When you talk about how green the grass is on the other side of the fence – and compare it to the greenness over here – you are measuring greenness by taking some average over many different blades on many small patches of land. In your eye, a similar process creates a sense of what the average intensity might be. (To get a feel for an average over a range of areas, see Exercise 4 at the end of Chapter 8.) The result is that the eye can create a response function or discharge rate $R(I)$ that is tailored to have its middle value (about 150) at the background level, with sensitivity maximized at nearby luminances. This response function is approximately logarithmic over more than one order of magnitude around this value, at which the discharge rate varies from about 50 to 250.

For a background of 10^1 cd/m^2, say, a one order of magnitude range around this number goes from $10^{1/2}$ to $10^{3/2}$ candles per square meter.

> *Suppose the background level of intensity is 10^1cd/m^2. Write a logarithmic model for $R(I)$ for $10^{1/2} \leq I \leq 10^{3/2}$, or about $3 \leq I \leq 30$.*

Thinking of the quantity $\log(I)$ as our input variable x, the discharge rate is a *linear* function of this x (see Appendix 6.2) passing through the points $(1/2, 50), (1, 150), (3/2, 250)$. Its slope is therefore 200, and we find

$$R(I) = -50 + 200x = -50 + 200 \log(I), \qquad 10^{1/2} \leq I \leq 10^{3/2}.$$

We have found a response function that should work well for intensities near 10 cd/m^2.

But this can't be the full story, since this function would imply $R(10^0) = -50$, an impossible discharge rate. So while the logarithmic function may be a good model for the response rate in a given range, beyond that range, something else must happen: beyond the low end of the range, the discharge rate approaches zero, and beyond the high end, the response saturates at about 300, and this happens at about one order of magnitude above the background. Above these intensities, the photoreceptor "maxes out," and all light looks equally bright. Some discharge rate functions are shown in Figure 9.2. This behavior is the physiological and mathematical reason behind the blinding effect we experienced upon emerging from that root cellar into bright sunlight: when the background intensity is lower, your eye maxes out earlier.

If you then change the background luminance – you turn off the lights; you step inside; you enter the cave – your eyes take some time to adjust to the new environment. This process resets the discharge rate, creating a new function $R(I)$ with response rate about 150 at the background intensity. The dust bunnies appear.

Solving the Riddle

The stars certainly don't suddenly hide behind the horizon when they see the Sun coming up, so where do they go?

We now know how to determine whether we can see a star in the sky. The answer will depend on the light intensity from the star and the background. So to decide, we will need to look up this information. We'll also have to decide how long we'll look up at the star to try to see it – say, 1 second.

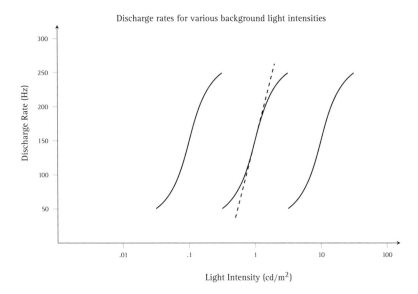

Discharge rates for various background light intensities

Figure 9.2 Discharge rate functions for three different background levels. The approximately logarithmic behavior is highlighted for the middle graph, which appears as a straight line due to the logarithmic scale of the x axis.

A quick search for starlight intensities reveals that the luminance of light from a typical star is about 0.001 candles per square meter.

Under what conditions will such a light source be visible to us?

Can you say?

Think of your response and jot it down.

Suppose there is some background intensity. I don't know what it is, so I'll give it a variable name, I_0 – reserving the unadorned I for the variable input. For the light source to be visible, we'll need an additional nerve firing in the presence of the star – and since we'll be looking for 1 second, we want the discharge rate to increase beyond the background rate by at least 1 per second. Now the key point:

for our discharge rate to be able to distinguish an additional 0.001 units of intensity from the background, we require $R(I + 0.001)$ to be at least one more than $R(I)$.

We could never have figured out this mathematical model without a good understanding of the biology.

We imagine that we are looking up at the sky, which is essentially uniform, except for the star in question. That is, we can assume that the intensity away from the star is the background intensity I_0 and the intensity of light with the addition of the starlight is $I_0 + 0.001$ (in cd/m^2). Now for values near a background intensity I_0, we can take the discharge rate to be logarithmic, as we have discussed (see Exercise 3). For the case of a background I_0, we find $R(I) \approx 150 + 200(\log(I) - \log(I_0))$, or equivalently, we have

$$R(I) \approx 150 + 200\log(I/I_0).$$

For our eye to register the star as distinct from the background, we require $R(I_0 + 0.001)$ to be at least 1 more than $R(I_0)$, and since $R(I_0) = 150$, this means at least 151. Plugging this intensity value into the discharge rate function, we find

$$150 + 200\log((I_0 + 0.001)/I_0) \geq 151,$$

or, after subtracting 150 from both sides and dividing by 200,

$$\log(1 + 0.001/I_0) \geq 0.005.$$

We want to find what this says about the background intensity I_0, so we would like to isolate that term. To get rid of the logarithm term, we use the fact that exponentiation is the inverse process of taking a logarithm. That is, $10^{\log x} = x$. So we exponentiate both sides to get rid of the logarithm and find $1 + 0.001/I_0 \geq 10^{.005} \approx 1.0116$, so $0.001/I_0 \geq 0.0116$, or $1/I_0 \geq 11.6$. This gives

$$I_0 \leq 1/11.6 \approx 1 \times 10^{-1},$$

where at the end we decide that we can work to just one significant digit.

What does this mean? Recall from the start of the chapter that this level of intensity corresponds to moonlight. So if the intensity is moonlight level or darker, we should be able to see the star.

Conclusion

Our conclusion is that a star should be visible in moonlight or darker. But above moonlight – for example, when the Evanston sky is bathed in the glow from Chicago's city lights – the light from stars may not be distinguishable from the background, and we have to travel to the woods for a starry sky.

Certainly, there are no stars visible in daytime. They are literally hiding in broad daylight!

Summary

Let us review how we were able to answer the question. First, we learned that the eye sensed light intensity by firing neurons at different rates. For us to sense the additional light of one star beyond its background, we needed the difference to be an amount that triggers at least one more firing. Empirically, the firing rate is a logarithmic function in the vicinity of the background intensity and varies by an amount of about 200 within an order of magnitude of intensity around the background. This determines how the firing rate changes, and we can figure out whether a star shining at a luminance of 0.001 triggers another firing – a question whose answer depends on the background. We found that a background intensity of moonlight (or lower) would allow us to see the stars, just as we might expect.

Exercises

1 Discuss the factors determining when we can see a daytime Moon.

2 Give a qualitative explanation of the optical illusion in the variable Hermann grid shown in Figure 9.3.

3 Write down a logarithmic (up to a constant) function $R(I)$, defined in the range 10^{-2} to 10^{-1}, whose values vary from 50 (low end) to 250 (high end). Show work! At what intensity I_0 is $R(I_0)$ equal to the average of these two, 150?

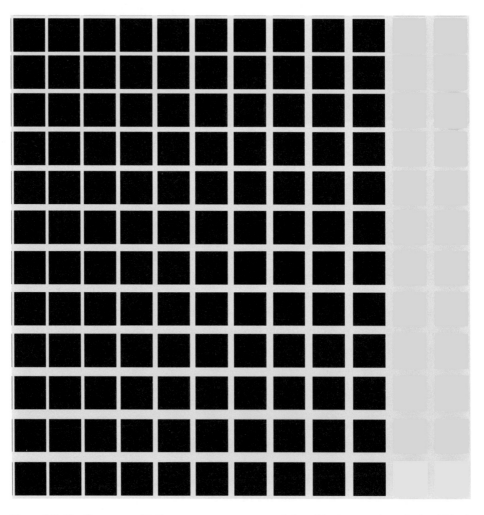

Figure 9.3 The Hermann grid. Focus your eyes on one of the white intersections. It should look white, though gray dots seem to occupy neighboring intersections. The strength of the effect varies depending on where in the grid you look.

Now suppose instead that I_0 is given. Write down a function $R(I)$ varying from 50 to 250 in an order of magnitude "centered" around I_0. (You should understand properly what is meant by "centered.")

4 After reading a book outside in sunlight (intensity $I_a = 10^4$ candles per square meter), you come indoors to a dark basement (intensity $I_b = 10^{-2}$ candles per square meter). Your eyes can't adjust to the change in background intensity instantly; rather, the background intensity detected by your eyes gradually decreases from 10^4 and approaches 10^{-2}. For the purposes of this exercise, assume that the detected background intensity I_0 as a function of time can be modeled as $I_0(t) = I_b + (I_a - I_b)e^{-10t}$, where t is measured in minutes. Assuming, as in Exercise 3, that your eyes can accurately detect light in a range of one order of magnitude centered around I_0, find how long it will take before you can see well enough in the dark. That is, how long before I_b is within the range of detection?

5 A supernova emits the same amount of energy in a few days (10^{-1} years) as an average star does in a lifetime of 10^9 years. We take this to mean that its intensity is 10^{10} times that of the Sun. How close would a supernova have to be in order to be visible during the daytime?

   Facts: the Sun is about 1.5×10^8 km away; blue sky has a light intensity of 4×10^3 candles per square meter, while the light intensity of the Sun observed directly is 1.6×10^9 candles per square meter.[2] Recall from Chapter 8 that intensity decreases proportionally to the square of distance.

6 One measure of a solution's acidity is the pH scale, which measures the concentration of hydronium ions (H_3O^+) in a solution. pH values follow a negative base-10 logarithmic scale; that is, a substance with a pH of 5 has 10^{-5} moles of hydronium ions per liter, and a substance with a pH of 8 has 10^{-8} hydronium ions per liter. Write a formula for the pH value as a function of the number of moles of hydronium ions per liter.

   pH values typically range from 0 to 14. Water has a pH of 7. pH values less than 7 are considered acidic, and values greater than 7 are considered basic. Suppose you want to make lemonade from concentrate. The recipe tells you to mix the contents of the 0.5 liter can with 2 liters of water. The concentrated lemonade has a pH of 2.3. What is the pH of the lemonade solution that you drink?

7 Think of a topic for a project related to this question.

---

[2] www.npl.co.uk/educate-explore/factsheets/light/light-(poster).

Projects

A Why is the sky blue? Why are sunsets red? How red is the Sun at dusk?

B Design an experiment that will test the eye's rate of adaptation to differing light intensities. What materials will you need? What kinds of data will you collect? Write a detailed procedure explaining the steps you will take. Now, do the experiment and prepare a presentation on your findings. What are you able to conclude?

C Can I hear a pin drop?

D Give a quantitative explanation of the optical illusion in the variable Hermann grid shown in Figure 9.3.

10

Should I Take This Drug for My Headache?

Appendix Skills: Probability, Statistics

OVERVIEW

To find out if a drug will be effective, we consider analyzing data from a mock clinical trial. This involves a fair amount of probability and statistics, so we spend some time shoring up our knowledge in these subjects. We can then weigh the drug's efficacy against its cost.

1 *We first recognize that the question is about evaluating the likelihood of a desired effect.*
2 *So we learn probability in some detail, including the normal distribution associated to probabilities of averaging large numbers of trials. We recall that its shape is determined by the standard deviation.*
3 *Next we create a mock clinical trial of a treated group and a placebo group, measuring how long it took headaches to go away for subjects in these groups.*
4 *We estimate the probability of the treatment group having their headaches go away in less time by chance, using the normal distribution. Since this probability is found to be very small, we reject the null hypothesis that the result came by chance, and conclude that the drug was responsible for the effect.*
5 *Finally, knowing that the drug was effective, we would still need to decide for ourselves if the cost of the drug was worth the slightly shorter amount of headache time.*

What to Consider When Considering

To take the drug or not? Seems like a simple question – just try it and see if it works, *bada bing, bada boom*. Really? You're just going to pop it in your mouth? Without even looking at the warning label? What if one of the possible side effects is immediate death? Would you still do it?

Of course, one of the possible side effects of crossing the road is immediate death, but we do it all the time because we estimate the risk as small. Even if you're not concerned about whatever the potential side effects are, the drug costs money, and if it only works 10% of the time, maybe you'd rather just wait out the headache. An economist would say you have to weigh the *risk* of spending your money on something that might not work against the *reward* of eliminating the pain.

We've *already* figured a lot out. We understand the question to mean, try to evaluate the likelihood of a desirable outcome (no pain) against the costs and possibilities of an undesirable result (continued pain or side effects). Then, based on this analysis, decide how to proceed.

When you take a drug, there is a chance that it will work and a chance it will do nothing, or worse. But just because there are two options doesn't mean that they are equally likely. These different outcomes occur with some *probability* – so before getting into a detailed analysis of the problem as stated above, it may be worth the time to consider some questions of probability in isolation before applying the methods to clinical drug trials. Starting with a simple example is a wise move: gotta learn how to crawl before we can walk.

What Are the Chances?

As with most problems in probability or statistics, the essence of this one can already be seen by considering coin tosses. Easy, right? Flip a coin – a *fair* coin.

> *What are the chances of getting heads?*
>
> *Think of your response and jot it down.*

Fair means that heads and tails are equally likely, so they both should occur about half the time. So the probability is one-half, or 50%.

Lots of probability in rolling dice, too. Roll a fair die.

> *What's the chance of getting more than four?*
>
> *Think of your response and jot it down.*

The answer is one-third. Out of the six possible equally likely outcomes, two of them (5 and 6) are more than four. Two out of six is one out of three, or 1/3.

Roll a pair of dice. Let's think about what the chances are of getting more than a ten.

> *How many possible dice totals are there?*
>
> *Think of your response and jot it down.*

The answer is 11, since the total can be $2, 3, \ldots, 11, 12$.

> *How many possible dice totals are more than ten?*
>
> *Think of your response and jot it down.*

The answer is 2, since only 11 and 12 are more than 10.

So there are 11 possible dice totals and 2 of them are more than 10.

> *Roll a pair of dice. What is the chance of getting more than a ten?*
>
> *Think of your response and jot it down.*

If you said 2/11, it's because you were set up. Though the dice may be fair, the possible totals when you roll more than one die are not equally likely! There is only one way to roll a 12, namely, boxcars (six and six). However, there two ways to roll an 11. To see why they are different, imagine that one of the dice is colored red, the other blue. Then you could get (Red 6, Blue 5) or (Red 5, Blue 6) to total 11. Each roll on the red die is equally likely, and each roll of the blue die is equally likely, so each pair (red roll, blue roll) is equally likely, and there are 6 × 6 such pairs. (Why?) If we make a chart of totals,

Blue Die							
6	7	8	9	10	11	12
5	6	7	8	9	10	11
4	5	6	7	8	9	10
3	4	5	6	7	8	9
2	3	4	5	6	7	8
1	2	3	4	5	6	7
	1	2	3	4	5	6

Red Die

we can see how many of the $6 \times 6 = 36$ possibilities result in a total of 2 (exactly 1, namely, snake eyes), a total of 3 (just 2), 4 (3), 5 (4), 6 (5), 7 (6), 8 (5), 9 (4), 10 (3), 11 (2), 12 (1). Three of the 36 possible pairs are rolls above 10, so the chance is 3/36 or 1/12.

When you run a drug trial to determine a drug's effectiveness or chance of working, you test many subjects. So we should consider probabilities after a process is repeated multiple times, such as the chance a particular succession of results occurs or the chance that the average result is such and such.

With dice, if I ask for the probability of rolling the particular sequence (Red 2, Blue 3), the answer is 1 in 36, since just one box in the chart above corresponds to this sequence. Another way of thinking about it is that there is a one in six chance of rolling Red 2, and then *given that we rolled a Red 2*, there is then a one in six chance of rolling a Blue 3. In other words, one-sixth of the total number of rolls should result in Red 2, and one-sixth *of those* will further result in Blue 3 – so the total fraction of (Red 2, Blue 3) is one-sixth of one-sixth, or $\frac{1}{6} \times \frac{1}{6} = \frac{1}{36}$. Note we multiply the probabilities.

Key point: we have assumed that the two events – the roll of the red and the role of the blue – are *independent,* meaning the fact that the red die came up 2 had no influence on whatever happens when we later roll the blue die. The same independence is used to model successive coin tosses: just because you happen to have flipped ten heads with a *fair* coin, it doesn't make it any more likely that tails will come up on the next throw. Many a gambling debt has been amassed from a false understanding of independence in probability: "Come on, seven! I'm due for a seven!"

If I roll three dice and ask for the chance that the sum is 6 or less, I can add up the chances that they add to 6, to 5, to 4, and to 3. Let's start with 6. We can enumerate all the ways that three numbers can sum to 6:

$(1,1,4), (1,2,3), (1,3,2), (1,4,1), (2,1,3), (2,2,2), (2,3,1), (3,1,2), (3,2,1), (4,1,1).$

There are ten ways total, and note that they are all equally likely, so the probability of rolling a 6 with three dice is the fraction of rolls represented by the list of ten above. We therefore divide 10 by the total number of rolls of three dice. The total number of rolls must be 6 times the total number of rolls of two dice, or $6 \times 6 \times 6 = 6^3$. The probably of rolling a sum of 6 is therefore $10/6^3 \approx 0.0463$.

Do the same with 5, 4 and 3. Write down your calculations.

In total, we find $10 + 6 + 3 + 1 = 20$ possible ways to sum to a 6, 5, 4 or 3, so the probability of rolling 6 or less with three dice is $20/6^3 \approx 0.0926$.

I admit it: this is tedious! It would be convenient to have a way to get these probabilities without going through this whole rigmarole. In fact, it's necessary. Drug trials are run with hundreds of participants, each one having a response akin to the roll of a die. Precise calculations would be a waste of time and effort: we just need a decent estimate.

Is This a Fair Coin?

We need to estimate probabilities to determine whether a positive result of a clinical trial might have happened merely by chance. To model this with dice or coins, we might flip a coin a bunch of times. Then, if we didn't wind up getting heads for about half of them, ask, *Is this so unlikely as to lead us to conclude that the coin is not fair, but rather weighted?*

Returning to our example with dice, a way to estimate the answer without going through the rigors is to assume that the average of the three dice follows a normal distribution. We will try to use this technique to estimate the probability of rolling 6 or less with three dice.

Remark 10.1 Forgotten normal distributions or never learned them? Go to Appendix 8.5!   ▲

Recall from Appendix 8.5 that the average of a large number of trials eventually follows a normal distribution. In our case, we have three trials, and although three is not large, we can treat it as a toy model. In this case, we are asking about measuring a *total* of 6 or less with three dice, or equivalently that the *average* of our three rolls is 2 or less (since six divided by three is two). The expected value or mean for rolling one die is 3.5. The nice thing about averaging rolls is that the mean doesn't change, so the mean for averaging three rolls is still 3.5. But we're interested in all the probabilities, not just the mean.

We flew past a subtle point. You could roll three dice and take their average (not the sum, rather the sum divided by three). Then you can roll them again and take *their* average. And again and again. These batch averages follow a probability distribution with their own mean and standard deviation. The mean is actually the same, but the standard deviation goes down: averaging a number of trials leaves you more likely to cluster around the mean. This point is demonstrated by Figure A9.

In fact, we know how the standard deviation behaves. If you know it for one die, then for *averaging* (not summing) two dice, the standard deviation is smaller by the factor $1/\sqrt{2}$. For averaging three dice, it is smaller by the factor $1/\sqrt{3}$. In this case, we calculated in Appendix 8.4 the standard deviation of rolling one die to be $\sqrt{35/12} \approx 1.7078$. So for three dice, we divide this quantity by $\sqrt{3}$ and get 0.986. The fact that the standard deviation gets smaller for batch averages is crucial.

We are interested in rolls of the die averaging 2 or less, where the mean is 3.5. So we are interested in recording a value of 1.5 below the mean.

How many standard deviations below the mean is 2?

Think of your response and jot it down.

Since one standard deviation is 0.986, the number 1.5 represents $1.5/0.986 \approx 1.52$ standard deviations. In statistics, this quantity – the number of standard deviations away from the mean – is called the Z-score. Here, for the value 2, we have $Z = -1.52$, because we are *below* the mean.

The cumulative distribution function Φ tells you the probability of measuring below a certain Z-score, so it is reasonable to expect that the answer to our question is simply $\Phi(-1.52)$, which we look up to be about 0.0642. This is actually not very close to the exact result, which was about 0.0926.

The reason we're off is because three is not a very large number of trials. If we had a very large number of trials, the probability of getting some particular average value like 2 would be minuscule. (In Figure A9, you can see that the height of each individual black dot is low, and these heights decrease as we increase the number of tosses.) So the difference between finding an average of "less than 2" versus "less than or equal to 2" would be of little consequence. Here, with so few trials, the probability of your average being precisely 2 is not insignificant, and we should interpret "less than *or equal* to 2" as something in between "less than 2" and "less than the next possible value greater than 2." In terms of summing the dice, the next highest possibility more than a sum of 6 (with its average of 2) is a sum of 7, which gives an average of $7/3 \approx 2.33$. This number is 1.17 below the mean, or $1.17/0.986 = 1.19$ standard deviations; its Z-score is thus -1.19.

Therefore, we will take the midpoint of the Z-scores of -1.52 and -1.17, which is -1.34. With this Z-score, we look up $\Phi(-1.35) = 0.0901$. This answer is actually a pretty good approximation of the exact result of about $0.0926!$[1]

What is the percentage error?

Think of your response and jot it down.

Answer: the percentage error is 2.7%. Thankfully, as we'll see, things will eventually get *easier* for a large number of trials, as the difference between some Z-score and the next one up will be small. (Readers may thus gloss over this fine point.)

We had a decent result with three dice, but did we just get lucky? Maybe so. As we confessed, three is a small number for employing statistical reasoning. To give some more confidence to this method, we should use a few more rolls of the dice. After all, the central limit theorem only says that *eventually* the distribution of the average of a bunch of trials approaches a normal one. What if we take 12 rolls of the dice and ask that the total is no more than 30? Then we are asking for the average to be less than or equal to $30/12 = 2.5$, and this is a distance of 1 below the mean. This time, the standard deviation is $\sqrt{35/12}/\sqrt{12} \approx 0.493$, so we are $1/0.493 \approx 2.0284$ standard deviations below the mean. The next number up gives 1.8593 standard deviations below the mean, and these two average to 1.944, giving an estimated probability of $\Phi(-1.944) \approx 0.0259$. The exact answer is $55268357/6^{12} \approx 0.0254$ – Less than 2% error. Quite close!

So our statistical methods become more reliable when we average more outcomes. This begs the question of how many you need in your sample to get a good approximation with the normal distribution. One rule of thumb used in statistics is that you need at least 30. Let's try an example with just coin tosses. Recall from Appendix 8.5 that if we take tails to be 0 and heads as $+1$, then a single fair coin toss has mean $1/2$ and standard deviation $1/2$. So if we toss 30 coins and take their average, the result will have mean 0.5 and standard deviation $0.5/\sqrt{30} \approx 0.091287$. So what if we toss 30 coins and ask the chance that we get no more than 12 heads? Twelve heads means 18 tails, and an average toss value of

$$\frac{1}{30}(18 \times 0 + 12 \times 1) = 2/5;$$

i.e., the toss average is the fraction of heads. So we're asking that the average is $2/5 = 0.4$ or less. How many standard deviations below the mean of 0.5 is this? It's $(-0.1)/0.091287 \approx -1.0954$. What's the next highest? That would be 13 heads and 17 tails, giving an average of $(13 - 17)/30 = -4/30 \approx -0.13333$, which is $-0.13333/(0.182574) = -0.73030$ standard deviations below the mean.

---

[1] We just constructed an argument and ran the analysis but found a discrepancy between prediction and observation (or calculation): the prediction using the Z-score for a roll average of 2 was way off. So we had to figure out an explanation for it, namely, the difference between < 2 and ≤ 2. This is a template for what good quantitative reasoners should do. In fact, when I was preparing this chapter, I overlooked this subtlety and had to confront the discrepancy this way for myself.

Averaging these two, we get $\frac{1}{2}(-1.0954 - 0.73030) \approx -0.91285$, so we look up $\Phi(-0.91285)$ and find it to be 0.18066. In this case, we can also compute the exact answer. As a decimal, it's about 0.18080. The approximation is off by a fraction of $0.00014/0.18080 = 0.00077$, so we're off by only 0.07%. Very, *very* good!

We are computing these probabilities in order to answer the question, is this a fair coin or is it weighted against heads? We just decided that a fair coin would have an 18% chance of coming up heads only 12 times or less, so our observation was unlikely, but not particularly rare – certainly not enough to justify a complaint to the Gaming Commission. To make any kind of sober analysis, we'd want to be more confident in our conclusion, setting a confidence level at 95%, say (or even higher for scientific reasoning). So our measurement has a probability (or *p*-value, in statistics lingo) of occurring by chance (the "null hypothesis"), and if p is found to be below 0.05, then we can assert with 95% certainty that something *besides* random chance caused our result. In the present case, we found $p = 0.18$, which is greater than 0.05, so we cannot reject the null hypothesis with confidence and thus cannot conclude that the coin is weighted. This type of argumentation is called *hypothesis testing*. It is discussed in Appendix 8.6 (see also Exercise 5). We next consider an application of this method to our headache problem.

A Statistical Model of Effectiveness

We can now figure out how to assess if a drug has some effect against a headache. We will run a trial with a large number of "random" people with headaches[2] and measure the effect.

> *If the headaches go away, does that show our drug is good?*
>
> *Think of your response and jot it down.*

It does not! For as you likely know, even untreated headaches go away at some point.

So what could a headache drug even do? I guess it could make the headache go away more quickly. Then we can decide to record *when* the headache goes away and see if the time is *reduced* after treatment with the drug. We have thus refined what we mean by an answer to the question, *does this drug work?* We will say that it works if it reliably shortens the time of your headache. Quantifying *reliably* will be the main point.

---

[2] Thinking about what we mean by "random" and how we run our trials opens a host of new questions. For example, we surely don't want concussed football players, since they may have nonstandard headaches that our drug was not designed to treat anyway. Also, how are you going to give them headaches? You probably can't do this in a laboratory setting (in a humane way), so you may have to ask them to administer the drug on their own at home – but then some will forget or misrecord the data. All sorts of things can – and will – go wrong. Designing proper protocols for drug trials can be an extremely complicated task!

We might compare the "feel-good" times (not a standard term!) that patients report *with* the drug to the feel-good times *without* the drug. More realistically – and to narrow our focus to the drug itself, and not the act of taking it or having it prescribed for us by a doctor – we would compare the effects of the drug being tested to a *fake* drug or "placebo," such as a sugar pill. The patients will not be told whether they are taking the drug or the placebo. This "blind" protocol helps eliminate confounding effects. For example, using a placebo helps discount the positive effects that might occur from a person simply feeling as though they are being treated, since these effects would be present for *both* groups, the treated group and the "control."[3] So the placebo should look just like the drug being tested.

We will try to determine if our drug works better than a placebo.

To keep the doctor's attitudes from affecting the outcome, too, the trial should be designed so that both patient *and* doctor are kept unaware of which pills are the real medicine, a "double blind" trial. We imagine running such a trial, comparing our drug to a placebo.

After running a very costly and lengthy experiment, we arrive at a range of feel-good times for the treated group receiving the drug and for the control group receiving the placebo – and both populations should have at least 30 people.[4]

Then what? Since we have sampled enough people, we can assume (see Appendix 8.5) that the distribution of times that we measure will be shaped like the standard bell curve of the normal distribution, centered at the average time, with the width of the bell described by the standard deviation. We will have two such curves, in fact: one for the placebo group and one for the treated group. We are testing the hypothesis that the drug is better than the placebo. So we will see if the average feel-good time is lower for the treated group. But that's not enough: we must then calculate the probability of getting that lower measurement purely by chance in the placebo group. If that chance (the *p*-value) is small enough, we will *reject* the possibility that the drug was no better than the placebo (the *null hypothesis*) in favor of the *alternative hypothesis* that the drug is better. We get to set our own threshold value (*significance level*) for this probability. Some standard values for the significance level are 5% and 1%. The lower the value is, the less of a chance of an error there is, so the more confident a conclusion you can make.

So far, this is just a strategy. To implement it in practice, we need the actual data. We'll pretend that we ran an experiment with 60 people, exactly 30 in the placebo and 30 in the group receiving treatment with the drug. For each person in each

---

[3] The "placebo" effect is real. If patients are told the placebo is a stimulant, some will show elevated heart rates and blood pressure. If they are told it is a sleeping pill, they'll exhibit the opposite response. Even "sham surgeries" can have a positive effect, though intentionally misleading a patient is of course completely unethical.

[4] There are other analyses to perform if the sample size is small for one or both populations, but in this introduction to the statistical methods of drug trials, we are considering a more favorable scenario.

group, we measured the feel-good time.[5] That is, we have two sets of data. Each data point is a time (in minutes, say) between taking the drug and "feeling good." In an actual experiment, you might want a less subjective measure of "feeling good." Defining such a measure poses another challenge to the design of your experiment. Maybe the time until the patient was willing to play his or her favorite song at some fixed, high-decibel volume? Perhaps. We'll content ourselves with sweeping the question under the rug.

Let's suppose we have the data, and the placebo group feel-good times look like this:

PLACEBO

28	23	18	21	21	24	22	18	21	24	$\bar{x}_p = 22$
13	25	22	19	21	25	21	20	19	19	$\sigma_p = 3.21$
21	23	26	28	25	19	25	24	20	25	

A computer is more useful here, but it's not too hard to see that the average is 22. If you'll pardon the sports metaphor, pretend these are scores on golf holes where the par is always 22. Then after the first hole, you'd be 6 over, then one more at the second hole so 7 over after 2, then $6, 5, 7, 7, 3, \ldots$. You finish at par. To calculate the variance, take the average of the square of your score above/below par on each role: here $6^2 = 36$ on the first, $1^2 = 1$ on the second, $(-4)^2 = 16$ on the third, etc. The variance is therefore $\frac{1}{30}(36 + 1 + 16 + \cdots)$, and we calculate this to be 10.33, so the standard deviation is the square root of this, or about 3.21. These numbers are indicated to the right of the data.

We now move to the treatment group:

TREATMENT

16	18	18	19	15	24	20	15	18	19	$\bar{x}_t = 18$
20	27	12	21	20	17	16	10	18	20	$\sigma_t = 3.84$
21	24	14	24	16	11	18	16	19	14	

It is easy to calculate that the mean is 18. The variance is about 14.73, so the standard deviation is about 3.84.

So to assess the drug's efficacy, we need to figure out the likelihood that the lower scores in the treatment group were just by chance. Was it just a fluke that the times went down, or can we say with confidence that they are statistically significant, i.e., attributable to the treatment? To answer this, we make the crucial insight that from the treatment group, we are only interested in the *single* measurement of their

---

[5] These are real people. Okay, they're not, but if they were, they might take the drug and not feel better even after several hours. What, then? Do we count the data point? We might – or we might have to decide that we will restrict our group to those who experience "nonpersistent" headaches. For our simple fictional experiment, we'll assume that there were no persistent headaches, or if there were, then they have already been weeded out.

average feel-good time, i.e., 18 minutes. But it is not a single headache vanishing in 18 minutes but the average of 30 times, and this carries more weight. Let us emphasize this point: having 30 people with an average of 18 minutes is a stronger indicator of drug efficacy than just having one person with an 18-minute headache. The chances of this lower average for so many people happening by fluke are slimmer – but how much slimmer?

We will begin by understanding the expected distribution of placebo headache times just for a *single* response, then use this to figure out how batches of 30 behave. So what is it? This is a delicate question, but here we make a *simplifying assumption*: we suppose that millions of similar experiments have been done with different drugs all over the world and that it is *known* that headache times of people in the placebo groups *always* follow a normal distribution with mean of 22 and a standard deviation of 3.21, as we observed in our sample. In reality, one hardly has perfect knowledge about the population, so one might make the assumption that the population behaves as observed, or employ more sophisticated statistical tools to deal with such cases.[6]

With this assumption, our perfect knowledge allows us to continue.

With our perfect knowledge of the population assumed, we have an expected normal distribution of headache times (measured in minutes) with average 22 and standard deviation 3.21. Okay, so now we have a treatment group of 30 with an average time of 18. What do we *expect* for an average of 30? Recall that the standard deviation of the average of items selected from the same distribution decreases by the square root of the number of items being averaged, while the mean does not change (see Box A4) – so we expect the average to follow a normal distribution with a mean of 22 still, and a standard deviation equal to $3.21/\sqrt{30} \approx 0.586$.

Now let's look at our single measurement from the group of 30 people: we found an average feel-good time of 18, which is a distance 4 from the norm. How many standard deviations is this? Since the standard deviation for the average of 30 is 0.586, our measurement is $4/0.586 \approx 6.83$ standard deviations below the norm. Our Z-score is therefore -6.83. What is the chance of measuring this average of 18 *or below*? That's the chance of getting this good a result or better from the placebo group. The value $\Phi(-6.83)$ computes the chance of measuring a quantity less than -6.83 in a normal distribution with standard deviation one, which is precisely equivalent to our question. We look it up: $\Phi(-6.83)$ is *incredibly small*, less than 1 in ten billion![7] Figure 10.1 shows the distribution of mean headache times for groups of 30 people from the placebo group.

---

[6] Our assumption makes life simpler for us but is unrealistic. A more sophisticated use of statistics would have us check the probability that these two data sets could come from the same distribution. If unlikely, then there would have to be some effect from the treatment.

[7] Anything beyond three standard deviations from the mean is rare: well below a 1% chance. Warning: in real life, your data probably won't be so clear-cut!

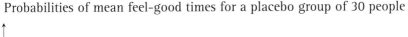

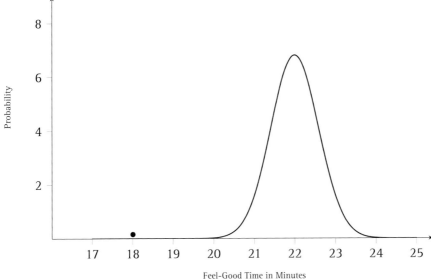

Figure 10.1 The *x*-axis is headache times in units of minutes. The dot represents the observed value (18 minutes) of the mean of 30 headache times in the treated group, a very improbable result under the null hypothesis.

Since the probability of generating an average feel-good time of 18 minutes for 30 people from a placebo group is so small, we reject the null hypothesis that this might have occurred by chance and conclude that the drug had an effect on the treated group.

Note that if we hadn't recognized that the distribution of averages of 30 headache times was different from the distribution for single headaches, we would have computed a single measurement of 18, which is -4 from the mean of 22, or $-4/3.21 = 1.246$ standard deviations. Now $\Phi(-1.26) = 0.11$, the would-be *p*-value, is larger than any reasonable significance level, so we would not have been able to come to a definitive conclusion about the drug.

Conclusion

Finally, we should return to the original problem. We have determined that the drug conclusively reduced the average feel-good times in a clinical trial. But even in that trial, not *every* treated subject had a lower time than 22, the average placebo-group member. So there is no *guarantee* that the drug will reduce our headache time from 22 to 18. And even if there were, we still would have to decide if we wanted it.

To answer the question quantitatively, let us suppose we know the cost of a package of ten caplets of the headache drug to be $25 – and for simplicity, suppose there is a very good chance that our headache times will be reduced on average by 4 minutes. So basically, that's $2.50 a pop for 4 minutes' less pain. Would you do it? It'll depend on your cash flow and your tolerance for pain. There is no right answer. This is an individual choice you must make, but after our analysis, it is an *informed* one!

Summary

No drug works all the time. If we give the same drug to headache sufferers, we will measure different effects. Before taking the drug, we cannot know for sure what will happen, but the evidence from experimental drug trials allows us to say something about the chance of success.

Chances and experiments are analyzed using probability and statistics, and the methods can be honed by considering coin tosses and dice. If all outcomes are equally likely, we can calculate the probability of an event occurring by computing the fraction of times it occurs among all possible outcomes. We can graph the different probabilities to get a probability distribution, with the average or mean falling somewhere in the "center." If we measure the average of a batch of outcomes – then do *this* many times – these batch averages themselves will have a distribution with the same mean, and if the batches are large enough, the distribution will be in a standard shape: the *normal* distribution, centered around the mean and falling off according to the standard deviation.

When clinical trials are run with large enough numbers of people, we can assume that the distribution of outcomes follows a normal distribution. We then compare a treated group with an untreated group and ask, could the difference in outcomes have come from chance, or were they far more likely to have occurred due to the drug treatment?

If the difference in the treated group (in our case, an average reduction of headache time by 4 minutes) is favorable and unlikely to have been a chance measurement from the untreated distribution, then we conclude that the treatment had an effect – and we are in a position to evaluate the question quantitatively: is the likely elimination of 4 minutes of headache pain worth the price of the drug?

Exercises

1   When you buy a bottle of pills,[8] you'll find a packet containing some of the data from experiments that the drug companies ran. Cite a relevant piece of data from an actual information packet. Explain its meaning and how it might affect one's choice to use the drug. (This exercise involves research.)

2   We noted some possible pitfalls in creating a clinical trial and gathering data. One was the possibility of a persistent headache. What other kinds of events might cause "bad" data points that will need to be rejected?

---

[8] This information can be obtained without purchase.

3 It takes me on average 15 minutes to get to my classroom. I leave my home every day at 10:43 AM to teach an 11:00 AM class. I arrive late for one in ten lectures. Assuming a normal distribution of travel times, what is the standard deviation?

4 An experimenter testing the time it takes mice to travel through a maze finds that the times follow a normal distribution with a mean of 42.8 seconds and a standard deviation of 4.3 seconds. What is the probability that a random sample of six mice would take more than an average of 45 seconds to complete the maze?

5 Estimate the chance of being struck by lightning this year. (This exercise involves research.)

6 Think of a topic for a project related to this question.

Projects

A At several points in our discussion, we noted the difficulties in designing a clinical trial. Think of a hypothesis or question you might want to explore by polling or experimentation and then describe a method to test it. Now *do the experiment*! Or, if that is really too cumbersome, gather *some* data and invent the rest as if you had done the full experiment – but then don't pass off your results as if the data were real. Analyze the data and draw conclusions about what the data say about your question.

B Create a game that incorporates elements of probability. Make the game as interesting and fun as you can. Determine the theoretical probabilities of all the outcomes in your game. Run trials that sample many plays of your game, either by hand or by computer. Do the experimental outcomes agree with your theoretical probabilities?

C Give a quantitative discussion of the issues around breast cancer screening, false positives, and false negatives. Discuss how longer survival rates may be due to early detection rather than improved care. Which stakeholders would have an interest in obscuring statistics one way or the other? Be sensitive to, yet do not fail to address, the many emotions and perspectives swirling around these issues.

Appendices

In these appendices we give lightning reviews of essential basic concepts, skipping many others. Because much of the treatment is dense, those with scant familiarity with these subjects should read *very* slowly, working calculations on the side with a pencil and paper while they read. Just skimming over the derivations is skipping the main point.

These appendices are pieces of mathematics. Why math? We want to assess the information we have and talk about it in a measured, quantitative way. We need numbers for that. Some information is not obviously "numerical," such as whether a person voted for Carter or Ford in 1976, but we can convert this into a number, say, by writing a 1 for Carter and a 0 for Ford. We will see many examples of representing information (temperatures, populations, vital signs, health, happiness, structural stability, electric charge, propensity for violence, pollutions levels, wealth, . . .) through numbers.

When two people get married, they often merge assets, and their combined wealth becomes the *sum* of their individual net worths. We will model such real-world quantities with numbers and then manipulate these numbers using the rules of arithmetic, such as addition. Now *your* net worth may be one number and *my* net worth may be another number. To talk about processes such as combining assets without nailing down exactly whom we are discussing or what some amount actually is, we should use a stand-in for the actual number, a *variable* such as x or χ or even $\aleph$. When we manipulate numbers-represented-by-letters, we use the laws of algebra, which is just arithmetic in disguise. If we use numbers to discuss future events or data involving many subjects, we encounter probabilities and statistics.

These appendices cover these topics and more. All of this math is relevant to the basic questions we pose throughout the book. Here we present the math in isolation, since it is crucial to understand these skills before tackling real-world problems.

1 Numeracy

Basic numeracy – a comfort and facility with numbers – is developed over time and experience, but a few pointers may help the reader understand how numbers tell us what they do. The skill is important, for while calculators are helpful for computations, they cannot give a feel for values and comparisons.

You know that fractions have a decimal representation, and many of us are more comfortable with the decimal expansion as it gives us an immediate sense of the amount. For example, you can easily see that 4/7 is less than 7/12 if you approximate them both as decimals (4/7 ≈ 0.57143 while 7/12 ≈ 0.58333), even though neither decimal expression is exact. As you likely know, fractions may have infinite decimal expansions (but they eventually repeat; in contrast, irrational numbers like $\sqrt{2}$, π and e have nonrepeating infinite expansions), so bear in mind that an *exact* decimal expression may be off limits – but this is typically of little practical consequence.

Let's talk our way through a few calculations. How much do you earn working a full-time job paying $30 per hour? Probably best to assume a 40-hour work week and a 50-week year. Even though these numbers may not be exact, they're probably close. Then working 50 weeks of 40 hours per week at $30 per hour means 30 × 40 × 50 dollars. This number has 3 zeroes at the end, following whatever 3 × 4 × 5 is. That latter calculation can be done by taking 4 × 5 first (20) and multiplying by 3 to get 60. Or, if you think 3 × 4 = 12 is easier to do first, maybe you know 12 × 5 = 60 automatically by recalling the face of a clock. And even if that wasn't so simple, you can fall back on this: five twelves is five tens plus five twos. In the end, we get $60,000/year. From this longhand derivation, we can even infer a rule for salary calculations: double the hourly rate and tack on three zeroes. (For more on units, see Section 5.)

> **The Calculator Question**
>
> Calculators are great for computing 78.53 ÷ 9.21, but they won't tell you why this is the right thing to compute. And if you get an answer of 852.7, they won't tell you that you missed the decimal point when you punched in 78.53. You need to have a sense of the outcome in advance, and your times tables will help get you there.
>
> Calculators are of limited use, and come with risks:
>
> - They are no substitute for numeracy.
> - They can't help you understand.
> - They don't always save time.
> - They won't tell you when you're wrong.
>
> The quick payoff of getting a number on the screen is a false crutch: that number is only likely to be correct if you have an understanding of your subject and a sense of what a reasonable answer might be. Better to work through your derivation and only pull out the calculator at the *end* to get a concrete figure.
>
> Even if calculators are permitted in math courses, they are unlikely to be of any real help.

You can use shortcuts to get a sense of a problem that you can't easily do exactly. Surely 29 × 31 can't be that far off from 30 × 30 = 900, since both numbers are close to 30. Even better, the first number is a bit smaller while the second number is a bit larger, so maybe some of the errors in the approximation get mitigated. This intuition is correct, and in fact the product equals 899. If you didn't need an exact answer, approximating 29×31 as 30×30 would get you pretty close.

1.4×7.3? Should be 7.3 plus 0.4 times 7.3, or 7.3 plus a bit less than half of 7.3, or 3-ish, so we could guess 10.3. (The actual answer is 10.22.) How about 243×552? You should parse this as something like 2 and a bit hundreds times 5 and a half-ish hundreds. Since 2 times five-and-a-half is 11, the answer might be something like 13 hundred hundreds, or 130,000. (The actual answer is 134,136. For more formal manipulations, see Section 5.2.)

If you tip 15%, you may easily calculate this as 10% plus 5%, or 10% plus half of that amount. Now 10% means one-tenth, so we just move the decimal one place to the left for this part of the calculation. So 15% of 48 would be 4.8 plus half of that, or 2.4, and these two add to 7.2. Or we could approximate 48 by 50 and then easily take $5 + 2.5 = 7.5$, the exact answer of 7.2 being a bit less. A 20% tip is easy: double 10%. What about 18%? This is 20% minus 2%, and since 2% is one-tenth of 20% we can think of this as 20% minus a tenth of that. So 18% of 77? Double 7.7 for 15.4 and then subtract a tenth of that, or 1.54. The result will be a bit less than 14, so you could guess 13.9. (The actual answer is 13.86.)

Similar games work for division – $1.42/3.7$? First read this as one-tenth of $14.2/3.7$, which is the same as $142/37$, an easier fraction to work with. Now since twice 35 is 70, and twice that 140, the ratio $142/37$ can't be much different from 4, maybe a bit less. Remembering to take a tenth, and a little bit less, leads us to 0.4 minus a bit, say 0.38 or 0.39? The actual answer is more like 0.3839, but our sense of numbers served as a good guide without calculation.

These tricks help you utilize the points of contact *that you already know* linking numbers to meanings and values, and enable you to get a feel for an answer even if finding it exactly is too time-consuming. For example, if you see a budget spreadsheet for your company, you might want to get a quick sense of the relative allocations of different departments, but without spending all the time to type in every figure you might want. The skill is also useful for recognizing calculator errors due to mistyped digits or forgotten decimal points.

Exercises

1 For each of the following, rank the four quantities from smallest to largest:

   a $6 + 4$ $6 - 4$ 6×4 $6 \div 4$

   b $10 + (-2)$ $10 - (-2)$ $10 \times (-2)$ $10 \div (-2)$

   c $5 + 0.1$ $5 - 0.1$ 5×0.1 $5 \div 0.1$

   d $(2 + 3) \times 4$ $2 + 3 \times 4$ $2 \times 3 + 4$ $2 \times 3 \times 4$

2 For each of the expressions below, indicate which of the numbers 1/100, 1/10, 1, 10, 100 is closest to the actual answer. Exact calculations should not be necessary, as a rough guess of the answer is all that is required.

a $36 \times .4$

b $889 \div 3.01$

c $2997 \div 3600$

d $314.1592653589 \times 0.03141592653589$

e $(334 + 356) \div 7890$

3 Which is larger? Exact calculations should not be necessary.

a 26.9×3.8 or 16.1×11.3

b $5.1 \div 1.7$ or $638 \div 399$

c 3% of 812 or 45% of 35

d 100×0.1 or $100 \div 0.1$

e 15×15 or 20×10

4 What is the value of $3 + 0.3 + 0.03 + 0.003 + 0.0003 + 0.00003 + \cdots$? Write the total as a fraction.

2 Arithmetic

A few concepts in arithmetic often trip people up later on in algebra, so let's solidify our knowledge by shoring up these foundations.

2.1 Distributivity

$$2 \times (4 + 5) = 2 \times 9 = 18$$

Likewise,

$$2 \times (4 + 5) = 2 \times 4 + 2 \times 5 = 8 + 10 = 18$$

After all, twice something means double it, so we can think of the calculation as

$$(4 + 5) + (4 + 5) = 4 + 4 + 5 + 5$$

and that's two fours plus two fives.[9] The same works if we take thrice instead of twice, so $3 \times (4 + 5) = 3 \times 4 + 3 \times 5$. This is called *distributivity*. Since we could make the same argument for any triple of numbers, we can say that no matter what numbers a, b, c represent we have

$$a(b + c) = ab + ac.$$

Thinking of the product ab as a the area of an a-by-b rectangle, we can easily visualize this relation as expressing the area of an a-by-$(b + c)$ rectangle in the following way:

---

[9] Note that we can rearrange the terms in the sum in any which way. This is due to *commutativity* and *associativity*.

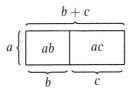

Since the rule applies for *any* numbers, that means we can use the same rule later on in algebra, where the numbers may be represented by letters or combinations thereof. The letters involved need not be named a, b, and c.

2.2 Fractions
Of course,

$$\frac{1}{2} + \frac{1}{2} \neq \frac{1}{4},$$

so you can't just add the denominators when you add fractions. If the denominators are the same, to add fractions, you add the numerators. Here we see two halves make a whole:

$$\frac{1}{2} + \frac{1}{2} = \frac{1+1}{2} = \frac{2}{2} = 1$$

If the denominators are *not* the same, we can *force* them to be by finding a common denominator:

$$\frac{2}{5} + \frac{3}{7} = \frac{2 \cdot 7}{5 \cdot 7} + \frac{3 \cdot 5}{7 \cdot 5} = \frac{14 + 15}{35} = \frac{19}{35}.$$

When we rewrite $\frac{2}{5}$ as $\frac{2 \cdot 7}{5 \cdot 7}$, we have multiplied it by $\frac{7}{7}$, which is 1, so we have changed nothing. Since the rule for adding fractions applies to all numbers, it applies when we use letters to represent numbers, too:

$$\frac{a}{b} + \frac{c}{d} = \frac{ad}{bd} + \frac{bc}{bd} = \frac{ad + bc}{bd}.$$

Here we multiplied the first fraction by $\frac{d}{d}$ and the second by $\frac{b}{b}$ before adding, to find a common denominator.

You know that $2 \times 3 = 6$ also means $3 = 6/2$. Likewise, we know $\frac{a}{b} \cdot \frac{b}{a} = \frac{ab}{ab} = 1$, so $\frac{b}{a} = \frac{1}{\frac{a}{b}}$. This simple observation means that dividing by $\frac{a}{b}$ is the same as multiplying by the reciprocal $\frac{b}{a}$.

2.3 Exponents
The expression $2^3 \cdot 2^5$ means $(2 \cdot 2 \cdot 2) \times (2 \cdot 2 \cdot 2 \cdot 2 \cdot 2)$. Since there are 3 factors of 2 in the first term and 5 factors of 2 in the second term, in total we have $3 + 5$ or 8 factors of 2, so[10]

$$2^3 \cdot 2^5 = 2^8.$$

---

[10] We freely use either $\times$ or $\cdot$ to denote multiplication, choosing mainly on aesthetic grounds. When a letter is involved, we also write $2a$ for $2 \cdot a$ or xy for $x \cdot y$, etc.

Likewise, $2^a \cdot 2^b = 2^{a+b}$ no matter what numbers a and b represent. However, $2^a \cdot 3^b$ cannot be simplified as such.

Note $2^3 \cdot 5^3$ means $(2 \cdot 2 \cdot 2) \times (5 \cdot 5 \cdot 5)$, which equals $(2 \cdot 5)^3$. We can write this rule as $a^r \cdot b^r = (ab)^r$. WARNING: Note how letters start to proliferate. If this ever gets confusing, replace the letters by numbers in a simple example like we did with 2,3, and 5, then work your way back to the general case. Use small numbers!

Now if $3^a \cdot 3^b = 3^{a+b}$, then $3^{\frac{1}{2}} \cdot 3^{\frac{1}{2}} = 3^1 = 3$, so whatever $3^{\frac{1}{2}}$ is, when you multiply it by itself, you get 3. That means $3^{\frac{1}{2}}$ is the square root of 3, or $3^{\frac{1}{2}} = \sqrt{3}$. Likewise, $3^{\frac{1}{3}} \cdot 3^{\frac{1}{3}} \cdot 3^{\frac{1}{3}} = 3$, so $3^{\frac{1}{3}} = \sqrt[3]{3}$. Now since $3^4 = 3^{0+4} = 3^0 \cdot 3^4$, this means $3^0 = 1$. Also, then $3^{-1+1} = 3^{-1} \cdot 3^1 = 3^0 = 1$, so $3^{-1} = 1/3$. Summarizing, and replacing our 3s by as, in general, we have

$$a^0 = 1, \qquad a^{-1} = \frac{1}{a}, \qquad a^{1/n} = \sqrt[n]{a}.$$

Now $(a^4)^3$ means $(a \cdot a \cdot a \cdot a) \times (a \cdot a \cdot a \cdot a) \times (a \cdot a \cdot a \cdot a)$, or $a^{4 \cdot 3}$ since there are three groups of four as. More generally,

$$(a^b)^c = a^{bc}.$$

This is consistent with what we already know. For example $(3^{\frac{1}{n}})^n = 3^{\frac{n}{n}} = 3$, so we rediscover that $3^{\frac{1}{n}}$ is the nth root of 3.

2.4 Percentages

Three percent, denoted 3%, means three out of a hundred, or 3/100 or 0.03. That's all there is to it. If a $150 dress is on sale for 15% off, then we must figure out what 15% of $150 is and subtract it from the price. Since 15% means 0.15 and "of" means "times," 15% of $150 means $0.15 \times \$150$, which is $22.50. (Quick tip: 15% is 10% plus half of 10%. Ten percent of $150 is $15, so when we add half again we get another $7.50, or $22.50 total.) Subtracting this amount gives the sale price of $122.50. Another method: since we have subtracted 15%, the result will be 85% of the original quantity, or 0.85 times it. Similarly, if you receive a 7% raise at work, your new salary will be 1.07 times your old one.

Some aspects of percentages can be tricky, but no more so than the same aspects of fractions. You might be asked the question in opposite order: a cell phone accessory is on sale at 25% off. The sale price is $45. What is the original price? In other words, 75% of what is $45? To get the answer,[11] we divide $45 by $75\% = 0.75 = 3/4$. Dividing by 3/4 means multiplying by 4/3, so $45 \times 4/3 = \$45 \times \frac{1}{3} \times 4 = \$15 \times 4 = \$60$. (The calculation was done "longhand" to reiterate some of the points about fractions made earlier.)

Another tricky point is illustrated by the following: start with 50. Subtract 20%. Now add 20%. Try it. (Don't think about it; just do it.) Where do you wind up? Not 50! Since 20% of 50 means $0.20 \times 50 = 10$, subtracting 20% from 50 yields 40.

---

[11] We could also pose this as an algebra question with the unknown representing the original price.

When we add 20% back, we will be adding 20% of this new quantity, 40 (not 50), so the number we add back will be different from 10, smaller. We compute: 20% of 40 is $0.2 \times 40 = 8$, so when we add it back, we get 48.

2.5 Logarithms

If you want to undo addition, you subtract. If you want to undo multiplication, you divide. If you want to undo exponentiation, you take the logarithm.

So if $10^3 = 1,000$, then $\log(1,000) = 3$. So what is $\log(1,000,000)$? Answer: 6. If you raise 10 to the 6th power you get 1,000,000. If you take the logarithm of 1,000,000 you get back the 6. The logarithm is the exponent (of 10), which gives the number. In other words,

$$\log(10^x) = x \quad \text{and} \quad 10^{\log(x)} = x.$$

The rules of exponents say $10^a \cdot 10^b = 10^{a+b}$; thus $\log(10^a \cdot 10^b) = a + b = \log(10^a) + \log(10^b)$. We learn that the log of the product is the *sum* of the logs:

(A1) $$\log(xy) = \log(x) + \log(y).$$

This is the flip side – since log is the undoing of exponentiation – of the rule that the exponential of the sum is the product of the exponentials. Note $(10^y)^x = 10^{xy}$, so taking log gives $\log((10^y)^x) = xy = x\log(10^y)$. Since 10^y could have been any positive number a, we learn that

$$\log(a^x) = x\log(a).$$

The quantity 10^x increases by a factor of ten each time x increases by 1, since $10^{x+1} = 10^x \cdot 10^1$. Likewise, $\log(y)$ increases by 1 each time y increases by a factor of 10.

There's not much special about 10 here. We could have used any number. The undoing of 2 to some power is written as the function $\log_2$, or "log-base-two." So $\log_2(2^5) = 5$, and the same rule (A1) applies for $\log_2$.

One special number, $e = 2.71828\ldots$, appears so frequently that $\log_e$ has its own symbol, ln, called the "natural log," and just as above, we have

(A2) $$\ln(xy) = \ln(x) + \ln(y),$$

as well as

$$\ln(e^x) = x = e^{\ln x}.$$

Likewise, $(e^y)^x = e^{xy}$, so ln gives

$$\ln(a^x) = x\ln(a).$$

There are a few more intricate rules. First,

$$\log_a(b)\log_b(c) = \log_a(c).$$

Why? Raise a to both sides. The right side gives c. The left side gives $a^{\log_a(b)\log_b(c)} = (a^{\log_a(b)})^{\log_b(c)}$, which is $b^{\log_b(c)}$, or c. In particular, this means $\log_a(b)\log_b(a) = \log_a(a) = 1$, so $\log_b(a) = 1/\log_a(b)$. These two together say

$$\log_b(c) = \log_a(c)/\log_a(b).$$

We can use this to find $\log_2(x)$ even if there is no base-two logarithm on our calculator. The rule says $\log_2(x) = \log_a(x)/\log_a(2)$, and we can pick a to be anything, such as e or 10. In particular, $\log_2(x) = \log(x)/\log(2) = \ln(x)/\ln(2)$.

2.6 Combinatorics

Counting things requires arithmetic at its most essential. If there are ten people and each has two legs, the number of legs is $10 \times 2 = 20$. If each leg has five toes, there are $10 \times 2 \times 5 = 100$ toes in all. Your combination lock has four cylinders, each with nine numbers, so the total number of combinations is $9 \times 9 \times 9 \times 9 = 9^4 = 6,561$.

We can also count choices. Suppose you are ordering from a sandwich chain and have to select a bread, meat, and cheese from the following menu:

Bread	Meat	Cheese
white	roast beef	provolone
wheat	chicken	cheddar
rye		Swiss
		muenster

Then there are three choices of breads, two choices of meat *for each* choice of bread, and four choices of cheese *for each* choice of bread and meat, giving a total of $3 \times 2 \times 4 = 24$ total sandwiches:

white-RB-prov, white-RB-ched, white-RB-Swiss, white-RB-muen   white-C-prov, white-C-ched, white-C-Swiss, white-C-muen
wheat-RB-prov, wheat-RB-ched, wheat-RB-Swiss, wheat-RB-muen   wheat-C-prov, wheat-C-ched, wheat-C-Swiss, wheat-C-muen
rye-RB-prov, rye-RB-ched, rye-RB-Swiss, rye-RB-muen   rye-C-prov, rye-C-ched, rye-C-Swiss, rye-C-muen

These kinds of counting problems are pretty simple: you use straight multiplication. But now what if the combination lock doesn't allow you to use the same number twice? Then there are nine choices for your first number, but only eight remaining choices of the second letter *for each* choice of first number, then seven for the third and six for the fourth, giving $9 \times 8 \times 7 \times 6 = 3,024$ total choices.

Consider for the moment a different problem: how many ways are there to arrange the letters ABCDE? This is a very similar question: five choices for the first letter (A, B, C, D, or E), four for the next, etc.: $5 \times 4 \times 3 \times 2 \times 1 = 120$. The *factorial* symbol 5! represents this descending product. So the answer is 5!.

Returning to the combination lock problem of listing **four** nonrepeating numbers from **nine**, note that the answer can be written with the factorial notation as

$$9 \times 8 \times 7 \times 6 = \frac{9 \times 8 \times 7 \times 6 \times 5 \times 4 \times 3 \times 2 \times 1}{5 \times 4 \times 3 \times 2 \times 1} = \frac{9!}{5!} = \frac{9!}{(9-4)!}.$$

Here the boldfaced numbers are hints at what to do in the general case.

The problem is trickier if the four different numbers form a "pick-four" lottery ticket instead of a combination lock, since in that case the ordering of the different numbers doesn't matter. So, for example, the four-tuples $(2,5,8,7)$ and $(8,2,5,7)$ represent the same lottery ticket, and the previous result of $9!/5!$ overcounts the total answer by treating these choices as distinct. The overcounting is by a factor equal to the number of ways there are of ordering four different numbers. This answer is $4!$. So we should divide by this amount, and the final answer is $9!/(5!\,4!)$, usually written $\binom{9}{5}$. This notation, read "nine choose four," is useful. If you want to count ways of rolling *nondoubles* in dice, that's two different numbers chosen from six, a kind of simpler version of our lottery ticket, giving $\binom{6}{2} = \frac{6!}{2!\,4!} = \frac{6 \times 5 \times 4 \times 3 \times 2 \times 1}{(2 \times 1) \cdot (4 \times 3 \times 2 \times 1)} = \frac{6 \times 5}{2 \times 1} = 15$. Indeed, these are the 15 rolls, written with the lower die first: $(1,2),(1,3),(1,4),(1,5),(1,6),(2,3),(2,4),(2,5),(2,6),(3,4),(3,5),(3,6),(4,5),(4,6),(5,6)$.

In general, the number of unordered collections of k elements chosen from among n is equal to $\binom{n}{k} = \frac{n!}{k!(n-k)!}$. Note $\binom{n}{k} = \binom{n}{n-k}$, since picking k things from n is the same as choosing which $n - k$ to leave behind.

Exercises

1 Circle the expression on the right that is equal to the expression on the left:

a $27 \times (5 + 13) = 27 \times 5 + 13 \quad 27 \times 5 + 27 \times 13 \qquad 27 \times 5 + 5 \times 13$

b $\frac{1}{2} + \frac{1}{4} \qquad = \qquad \frac{3}{4} \qquad\qquad \frac{2}{6} \qquad\qquad \frac{1}{8}$

c $x^a y^b \qquad = \qquad (xy)^{a+b} \qquad (xy)^{ab} \qquad x^a y^b$ (cannot simplify)

d $\log(\frac{x}{y}) \qquad = \qquad \frac{\log x}{\log y} \qquad \log x - \log y \qquad \log(\frac{x}{y})$ (cannot simplify)

2 Which of these represents the number of ways to order six different paintings on a wall?

$$6 \times 6 \qquad 6^6 \qquad 6! \qquad 6$$

3 Why is $1/(1/3) = 3$? Try to explain without formulas or numbers.

4 One measure of air quality used by the US Environmental Protection Agency is the amount of carbon monoxide in the air, measured in parts per million (ppm). So, for example, 6 ppm means for each million molecules in the air, about 6 of them (on average) will be carbon monoxide molecules, or 0.000006, i.e., 0.0006%.

   The EPA reports[12] a 61% drop in the concentration of carbon monoxide over 16 years. If we assume that each year the carbon monoxide dropped by

---

[12] www.epa.gov/air-trends/carbon-monoxide-trends#conat.

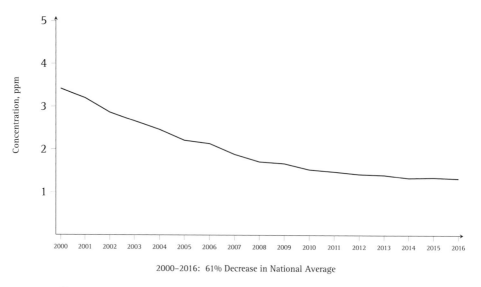

2000–2016: 61% Decrease in National Average

Figure A1 CO air quality, 2000–2016 (national average based on 155 sites)

a fixed percentage (this means a fixed percentage of what it was *at the start of that year*), what would that percentage be? Why is the naïve guess wrong?

5 My meal at a restaurant costs $100. There is a 10% tax, and I have a coupon for 10% off. What do I pay? Now redo your calculation applying the tax and the coupon in reverse order of what you just did. What did you find? Why?

6 If a chicken-and-a-half lays an egg-and-a-half in a day-and-a-half, how long would it take two chickens to lay a dozen eggs?

7 You are the editor of National News, a start-up website. Your chief health reporter has inside information about health care premiums. She tells you that there are two major health care providers that cover the vast majority of insured Americans, Unicare and Americare. Unicare covers about three times as many people as Americare. Anyway, the two will be raising their premiums by 5% and 1% next year, respectively. An average increase of just 3% would be a major news break, since this would be below the inflation rate of 4%. Can you write the headline "Healthcare premiums to rise just 3%. Below inflation rate for first time in decades"?

8 You are CEO of a construction company. Your Emerging Markets division has determined that there will be a major growth expansion in the state of Washington. They have analyzed that you will earn a dime of profit each year on every dollar spent on operations, over the next five years, but it will cost $100 million just to set up shop. (After that, you will begin to earn money.) The bank is offering you a $200 million loan at 5% interest per year for five years. Should you take it? What is the minimum five-year loan at 5% interest

per that you should accept? (Interest is the amount, in percentage, that you have to pay each year to get the money from the bank. They don't just *give* it away!)

9 When I mix one cup of 1% milk with two cups of 4% milk, what percentage milk will I get?

10 What is $\log\left[\left(\frac{5}{6} + \frac{2}{3}\right)^2 \cdot \left(\frac{1}{4} + \frac{7}{36}\right)\right]$?

11 At a make-your-own-pizza restaurant you can choose a single cheese from among three different kinds, and a single topping from among four different choices. How many kinds of pizzas can you make? How many if feta is required to pair with spinach (but not vice versa)? How many if we allow multiple toppings? Now answer the last question where the options of no cheese and no topping are permitted.

12 You begin at the point $(-4, -4)$ on a coordinate plane. If you are only allowed to move in one-unit steps either to the right or up, how many different paths are there to the point $(4,4)$? What if you must travel through the point $(1,0)$?

3 Algebra

"I can't remember her name but she's outgoing and she plays the violin." "Oh, must be Lydia." Here we have an unknown (who), and we figure it out based on properties it enjoys (outgoing, violin playing). In math, we use algebra to deduce what an unknown quantity is based on relationships (equations) it obeys.

We have just seen in Appendix 2 that letters can be used to represent numbers and can be manipulated in the same way as numbers. The rules are the same. That's the first key lesson of algebra, and we will use it repeatedly in this section. First, we must come to terms with the need to use letters instead of numbers when the value of a quantity is not known.

3.1 The Need for Variables; the Value of Formulas; the Case for Abstraction

In elementary school, you meet questions like $3 + \square = 7$ and are asked to fill in the square. Then maybe $8 + \square = 13$. After a few of these, you soon learn that the number in the square should be the number on the right minus the number on the left. Soon you do a few problems like $3 \times \square - 1 = 11$ or $2 \times \square = \square + 3$, but it is years before you encounter the equations $3 + x = 7, 8 + x = 13, 3x - 1 = 11, 2x = x + 3$. And yet they're all the same thing: $\square$ has turned into x. In these problems, you find the value of an unknown quantity based on a relationship (equation) you know that it satisfies.

When discussing a quantity without knowing its value, it is convenient to *assign a letter variable*, such as x, to represent the quantity. In real-life applications, you may want to determine some quantity of interest (such as how much they charged you for that cappuccino) based on information you know (such as your credit card receipt, which included tax and a 20% tip). The variable (such as x, but you may use whatever name you like) is the quantity of interest; the equation is the relationship it enjoys. Algebra is the toolbox which allows you to "solve for x."

In fact, with slightly more sophistication, we notice that *all* of those grade school problems are of a single form,

(A3) $ax + b = cx + d,$

with different values of a,b,c,d – so at some point you can invest yourself into solving this *one* problem with these *variable* entries.

Note the dual roles that variables play: (1) they can be a placeholder "box" for an unknown quantity, which we may be able to determine later (e.g., as above when we have an equation like $3 + x = 7$); or (2) they work in formulas when replaced by *any* number, allowing a single formula to address an infinite number of possibilities, as in the distributive law $a(b + c) = ab + ac$.

We can solve Equation (A3): if $a \neq c$, then $x = \frac{d-b}{a-c}$, whereas if $a = c$, then it depends: if also $b = d$ then any value of x solves the equation, while if $b \neq d$, there are no solutions. This result, a *single solution*, can be applied to the hundreds and hundreds of problems you solved as a youngster, and many that you will solve in this course! That is the value of deriving a formula: after adding just a small bit of abstraction (more letters) and sophistication (recognizing the universality of past problems), you get an answer to everything.

In this book, and in other real applications, we might

What Is an Equation?

Every equation looks like

$$A = B$$

where A and B are two expressions. So? Who cares? Well the key is that A might be hard and B might be easy. The equation tells you a simpler route. Outside of math, we might learn that *pulchritude = beauty*, replacing a hard word with an easy one. Note that it would make no sense if the parts of speech were mixed up, like if someone claims that *perseverance = about*. One is a noun and the other a preposition. Apples and oranges. (Well, at least those two are both fruit!)

Understand what *kind* of object is on the left of an equation and what kind of object is on the right. If you see

$$1 + 2 + 3 + \cdots N = N(N + 1)/2,$$

you must recognize that both sides are *functions* (see Appendix 6), since they both depend on N (unless N is known to be a particular value like 7, in which case both sides equal 28). Furthermore, on the left side, it must be the case that N is a *positive integer*; otherwise, the expression does not make sense. So this equation is of the form $A = B$, where A and B are functions of a positive integer N.

The utility of the equation comes from the fact that A is hard if $N = 200$, say, whereas B is simply $200(201)/2 = 20,100$.

consider ten different problems where some quantities (a loan balance, a population of flies, an amount of a radioactive isotope, a temperature of a baked potato, etc.) satisfy the same kind of relationships (changing by a rate proportional to their value), but with different parameters. Algebra allows us not only to assign variables (or letters) to represent the quantities of interest but *also* the different parameters (interest rates, birth rates, decay rates, heat transfer coefficients, etc.). In this way, we can tackle the *universal* problem and apply it in a variety of settings. As we shall later see, the exponential function captures the behavior of *all* of these systems. By working with exponential functions such as ab^t, with variable parameters (a,b), we will be able to address many applications at once.

Remark A2 Let's face it: too many letters make it hard to concentrate. When you are reading math or formulas and the letters get too much, either try to picture them as numbers or actually replace them with numbers as you follow along with pen and pad. Then go back later and see that the numbers didn't really matter. For example, when we solved $3 + \square = 7$ by subtracting 3 from both sides to see $\square = 7 - 3$, it didn't matter that the number on the left was an actual 3 or that the unknown was drawn as a $\square$ instead of an x: so we understand that the more universal problem $b + x = d$ has the solution $x = d - b$. ▲

3.2 Combining Like Terms

Suppose someone gives you three boxes of chocolates from a store. Each contains the same number of chocolates, but you don't know what number that is. Let's call it x. Then you have $3x$ chocolates in your possession. If another friend arrives with two more identical boxes, then you have $2x$ more, or $3x + 2x = 5x$ total, since you have five boxes altogether. This is *combining like terms* (both $3x$ and $2x$ involve just x). The same reasoning works if we think of x as representing a *type* of item like "apples" or a *unit* like "kilograms." You can't add kilograms and pounds until you convert one to another – see Appendix 5. In fact, combining like terms is just using the distributive law, since $(3 + 2)x = 3x + 2x$, and the first term is $5x$ since we can evaluate the term in parentheses before multiplying.

The terms you combine must be alike: $3x + 2y$ cannot be simplified. However, the same reasoning as above shows $3n^2 + 2n^2 = 5n^2$. This calculation would be relevant to the example above if you knew the chocolates were arranged in a square array and set n to be the (unknown) number of chocolates along the square's edge.

3.3 FOIL

Consider an expression like $(x+2)(x-3)$. We can parse this as a single term $(x+2)$ times the sum of two terms, x and -3.[13] Using the distributive law $a(b + c)$, with a set to $x + 2$, b set to x and c set to -3, we know

$$(x + 2)(x - 3) = (x + 2)x + (x + 2)(-3).$$

[13] It is ugly to write $x + (-3)$, so we just write $x - 3$, but these two quantities mean the same thing.

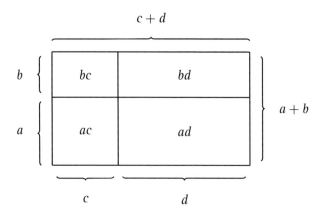

Figure A2 Visual demonstration of FOIL: $(a + b)(c + d) = ac + ad + bc + bd$.

At this point we run distributivity for each of the two terms on the right. Each is a product of one term with the sum of two others, and you can write the singleton on the left if you like. So we have

$$(x + 2)x + (x + 2)(-3) = (x^2 + 2x) + (-3x - 6) = x^2 - x - 6.$$

Note at the end that we combined like terms.

The general case may be easier to consider since the repeated appearance of x in the example above sometimes leads to confusion. If we write $(a + b)(c + d)$ and apply the same reasoning, we get

$$(a + b)(c + d) = ac + ad + bc + bd,$$

and this is often called "FOIL" for the mnemonic "Firsts-Outers-Inners-Lasts."

In Appendix 1, we saw that 29×31 was close to 900, and if we write $29 = 30 - 1$ and $31 = 30 + 1$, then FOIL tells us $(30 - 1)(30 + 1) = 900 + 30 - 30 - 1 = 900 - 1$.

For example, $(2x + 3x^2)(4x - 5)$ has four terms by FOIL, $8x^2 - 10x + 12x^3 - 15x^2$, but we can then combine like terms to get $-7x^2 - 10x + 12x^3$, or $12x^3 - 7x^2 - 10x$ if we follow conventions by writing the terms of a polynomial in descending order of the exponents.

Let's try $(x + a)^3$. Whatever it is, it's $(x + a)(x + a)^2$, and the latter term is $x^2 + xa + xa + a^2$ by FOIL, or $x^2 + 2xa + a^2$. So we must calculate $(x + a)(x^2 + 2xa + a^2)$. To approach this with FOIL, let's group $(2xa + a^2)$ together and write the calculation as $(x + a)[x^2 + (2xa + a^2)]$. Now FOIL says this is

$$x \cdot x^2 + x \cdot (2xa + a^2) + a \cdot x^2 + a \cdot (2xa + a^2).$$

Distributing twice gives $x^2 + 2x^2a + a^2x + ax^2 + 2xa^2 + a^3$, or $x^2 + 3x^2a + 3xa^2 + a^3$. We could have gotten here a bit faster if we'd first derived a more general distributive law $a(b + c + d) = ab + ac + ad$. This and its cousins with more terms in the parentheses can all be reduced (can you see how?) to the original case.

In fact, if we think of $(x+a)^3$ as $(x+a)(x+a)(x+a)$, the answer is the sum of all choices of multiplying one from the first group, one from the second, and one from the third. There are $2 \times 2 \times 2 = 8$ such choices, which is why the coefficients 1, 3, 3, 1 add up to 8. (The 3s appear because some of the choices give the same result, such as an x from the first group and as from the others, or an x from the third group and as from the first two.)

3.4 Solving for x; Solving for x and y

If we know $3x+77 = 146$, to solve for x we need to isolate it. To get rid of the "+77," we need to do the "opposite" of adding 77, which is subtracting it. Now we can't just subtract 77 from the left side; to keep both sides equal, we must do the same to the right side, too. Subtracting 77 from both sides gives $3x + 77 - 77 = 146 - 77$, or $3x = 69$. Now you may already see the answer, but let's continue the process of isolating x. Since x has been multiplied by 3, we need to do the "opposite," which is dividing by 3 – and do this to both sides. This yields $3x/3 = 69/3$. Doing the division, we get $x = 23$.

If there are xs on both sides of the equal sign, we need to gather them together first:

$(x + 7) - 15 = 6(x - 4) - 13$	First distribute on both sides.
$2x + 14 - 15 = 6x - 24 - 13$	Now simplify.
$2x - 1 = 6x - 37$	Add 1 to both sides.
$2x = 6x - 36$	Subtract $6x$ from both sides.
$-4x = -36$	Multiply both sides by -1 (to make it look nicer).
$4x = 36$	Divide by 4 on both sides.
$x = 9$	

Sometimes you have two variables (say x and y) and two equations:

$$3y - x = 4, \quad 2x + 3y = 19.$$

Adding x on both sides of the first equation gives $3y = x + 4$, and subtracting 4 to isolate x gives $x = 3y - 4$. Now look at the second equation, knowing $x = 3y - 4$. It says

$$2(3y - 4) + 3y = 19.$$

Distributing on the left side gives

$$6y - 8 + 3y = 19,$$

or $9y - 8 = 19$, and adding 8 gives $9y = 27$, which tells us (upon dividing by 9) that $y = 3$. Now we go back to the solution of the first equation, which said $x = 3y - 4$. This means $x = 3 \cdot 3 - 4 = 5$. So $x = 5$ and $y = 3$. Finally, we go back and check that these *do* solve the original equations. Check!

Summarizing, to solve two "simultaneous" equations with two variables (say, x and y), first *pretend* one of the variables (say, y) is a constant and solve the first equation. So you'll have $x =$ "something." Now replace all instances of x by that "something" in the second equation. Since "something" will have ys in it, but no x, the second equation will have only ys. Solve this for y, then replace y by that solution in "something" to find out what x is.

3.5 Word Problems

For many of us, algebra means word problems. In some sense, this book is about really complicated word problems. Most *simple* word problems can be approached with a few basic steps.

Some data will be known, some will be unknown, and some information will be needed to determine the known from the unknown. Often, these pieces of data will be represented by numbers, and for shorthand (or because we don't yet know the numbers) we call the numbers by some judiciously chosen letter. Sometimes reasoning with letters can be too abstract, so you are free to pretend they are numbers, preferably easy numbers. Sometimes I write the number 2 and then kind of massage it until it looks like the letter R, say.

After that, we want to be sure to translate the information from the words to equations. Convert each piece of information into a mathematical expression until you've gone through the whole story. In other words, keep squeezing the orange until you're sure there's no more juice.

Probably best to proceed by example.

A piece in the *New York Times* on December 28, 2014, reported the following about the movie *The Interview*:[14]

> *The Interview* generated roughly $15 million in online sales and rentals during its first four days of availability, Sony Pictures said on Sunday.
>
> Sony did not say how much of that total represented $6 digital rentals versus $15 sales. The studio said there were about two million transactions over all.

The reporter (whom we shall not name) failed to see this as an algebra problem.

What is the information that we want to know? The number of rentals – let's call this number R – and the number of sales, which we'll call S. What information do we know? There were 2,000,000 total transactions. That's the last sentence. We can convert that to a numerical expression: $R + S = 2{,}000{,}000$. So much for the last sentence. We can cross that off. What else do we know? There were $15,000,000 in total online sales and rentals (the first paragraph), and we know the cost of a rental and a sale (the remaining sentence). Note that the information that these figures were for the first four days and the fact that the studio is Sony Pictures are not relevant to question we are answering.

---

[14] www.nytimes.com/2014/12/29/business/media/the-interview-comes-to-itunes-store.html.

Now let's convert the rest of what we have to a numerical expression. To make the analysis simple, suppose there were just two rentals. Then we'd have $\$12 = 2 \times \$6$ from rentals; following this pattern, with R rentals we get $6R$ dollars. Likewise, with S sales we get $15S$ dollars. Thus the totals are $6R + 15S$, and they add to $15,000,000$ dollars. Summarizing:

$$R + S = 2,000,000,$$
$$6R + 15S = 15,000,000.$$

We have reduced the word problem to a purely algebraic system of equations, as in the previous section. The first equation can be solved for $R = 2,000,000 - S$ (we could alternatively solve for S). and we can use this value of R in the second equation:

> **Anatomy of a Word Problem**
>
> Let's walk through a typical word problem:
>
> > Henry wants to invite his friends to a chicken dinner. He has 20 drumsticks and a dozen ears of corn, and wants to use them all but be sure that all partygoers share equally. What is the largest number of friends he can invite?
>
> *Stay calm!* We just need to walk through to convert this to a simple math problem. These few questions work for many examples. *What kind of answer are we looking for?* A number here – the number of guests. *What is the relevant information?* All partygoers must share all food items equally. *What is irrelevant?* The name Henry, the types of food, the storyline. *How is this math?* Here's the crux: to share equally, the number of partygoers must divide 20 and 12 – and we want the largest such number. *Solve.* The largest number of partygoers is 4; therefore, since Henry himself is part of the party, he can invite three guests. *Answer:* 3.

$$6(2,000,000 - S) + 15S = 15,000,000.$$

Using distributivity and combining like terms, we get

$$9S = 3,000,000.$$

Dividing by 9 on both sides solves for S, but we find that S is a fractional value. We must recall that this is just a mathematical model for a reality, in which there is no such thing as a fractional sale of a video! So we round to the nearest whole number (the article only said "about" 2,000,000 transactions total, so these are all approximations anyway), and we get

$$S = 333,333, \qquad R = 2,000,000 - S = 1,666,667.$$

Because of that "about," we should probably round these to the nearest ten or hundred thousand – see Section 5 – but as a matter of algebra, the answer was not hard to find.

Some quick tips for attacking word problems: "and" is plus, "of" is times, "per" is divided by, "percent" is divided by one hundred.

Exercises

1 Circle the expression on the right that is equal to the expression on the left:

   a $4x + 7y$ $=$ $11(x + y)$ $11xy$ cannot combine terms

   b $4x \times 7y$ $=$ $28(x + y)$ $28xy$ cannot combine terms

   c $(x + 1)(y + 2)$ $=$ $xy + x + y + 2$ $xy + 2$ $x^2 + 3xy + 2$

   d $(x + 6)(x - 1)$ $=$ $x^2 - 6$ $2x - 5$ $x^2 + 5x - 6$

2 Evaluate the expression $x^3 - xy^2 + \frac{2x+y}{x} - 1$ for each of the following values of x and y. Which pair is a solution to the equation $x^3 - xy^2 + \frac{2x+y}{x} - 1 = 0$?

   a $x = 2, y = 0$

   b $(x,y) = (1, -1)$

   c $(x,y) = (1,2)$

   d $(x,y) = (a,3a)$

3 Using FOIL, show $a(b + c) = ab + ac$.

   Hint: $a = a + 0$.

4 Prove that $(x + y)(x - y) = x^2 - y^2$.

5 What if there are three terms? Prove $(a + b + c)(d + e + f) = ad + ae + af + bd + be + bf + cd + ce + cf$. Can you see what picture to draw?

6 Calculate $(x + a)^4$. Verify that the sum of the coefficients of the terms is 16.

7 Simplify $\frac{(x^2 y)^3}{x^5 y^{\frac{1}{3}}}$ Check your answer when $x = 3$ and $y = 8$.

8 I signed up for a frequent flier credit card that comes with 20,000 free miles and gives three miles for each dollar spent. If I spend x dollars, how many miles will I have? Use your answer and algebra to solve the following: if I have 38,000 miles, how much money did I spend with the card?

9 Propane is a clean-burning gas. A "gas" is a bunch of molecules floating around. The chemical symbol of propane is C_3H_8, which means that each propane molecule contains 3 carbon atoms and 8 hydrogen atoms stuck together somehow. What we mean by "burning" is a chemical reaction resulting from a combination with oxygen (most substances require heat to induce that chemical reaction). When propane combines with oxygen gas (symbol O_2 – the molecules are formed by two atoms stuck together), after burning, the resulting substances are carbon dioxide (CO_2) and water (H_2O) – this is what I mean by "clean." (Note that I have defined the terms of my sentence: "propane," "clean," "burning," "gas.")

What does a "single" reaction look like? It looks something like

$$C_3H_8 + O_2 \longrightarrow CO_2 + H_2O.$$

But that can't be right! There are three carbon atoms on the left and only one on the right, as well as other problems. So we need to adjust the CO_2 on the right a bit. But this affects the left. What we should do is say there are "a" molecules of propane (so $3a$ carbon atoms, for example) and b molecules of oxygen, producing c molecules of carbon dioxide and d molecules of water, and try to figure out what $a, b, c,$ and d are.

Hint: (1) Write down three equations for $a, b, c, d,$ one for each element involved. (2) Find some solution of these equations. (3) Find the minimal solution of these equations in which all $a, b, c,$ and d are natural numbers

Post mortem: Did you do this by guessing the answer? Can you be more systematic? Had I assigned you a harder problem, you wouldn't have been able to guess. (So why didn't I? Well, you also wouldn't have been as happy.) Try again without guessing.

10 Glucose ($C_6H_{12}O_6$) burns just as cleanly as propane. Do the same exercise as above to determine the numbers of molecules involved in the most basic such chemical reaction.

11 A piano is "well-tempered," meaning that the frequencies of successive half-notes differ by a constant factor. (Can you understand what this means? What does it mean?) If two notes differ by an octave, i.e., by 12 half-steps, then their frequencies differ by a factor of two. If this is so, then what is the frequency factor between C and G in the same octave? On other instruments, this interval is supposed to be a factor of 1.5. You should find that the piano is a bit off. By what percentage is the piano's factor off? How did you arrive at your number?

12 Calculate $(a+b)(c+d)(e+f)$. Can you draw a picture that explains the result?

4 Geometry

Many basic calculations require a bit of geometry. An environmental engineer (or you) may be interested in the amount of a particular pollutant in the Earth's stratosphere. This will surely involve the surface area of a sphere, at some point. Here we review the very basics.

The basic rule of geometry is the Pythagorean theorem: in a right triangle, if we call the length c of the hypotenuse (long side) and call the other side lengths a and b, then these values are related: $c^2 = a^2 + b^2$. This lets you calculate all sorts of perimeters of polygons, after breaking them down into (right) triangles.

As for round things, we'll start with circles. First, recall that the diameter of a circle is the distance across, twice the radius: $d = 2r$. Second, remember that the number π is *defined* to be the ratio of the circumference of a circle to its diameter, or in other words, the circumference is π times the diameter, i.e., πd or $2\pi r$. Now if an arc traverses a fraction of a circle, then its length is that same fraction of $2\pi r$. For instance, a semicircle has length $\pi r = \frac{1}{2}(2\pi r)$. If the arc is measured in degrees, the fraction of the whole is the number of degrees divided by 360.

As an example, if you want to fence in a plot of land shaped like a 30 degree sector of a circle with radius 90 feet, we can determine how much fencing you will need: there are two straight fences of 90 feet, and a curved one which is $\frac{30}{360} = \frac{1}{12}$ of the circumference of the circle, so $\frac{1}{12} \cdot (2\pi)(90)$ feet. The total is $180 + \frac{1}{12}(180\pi)$ feet, or $180 + 15\pi \approx 230$ feet of fencing (to two significant digits – see Appendix 5.3).

In your work as an environmental engineer, you may want to estimate the volume of the stratosphere, which occupies space at a distance between 10 and 20 miles above the Earth's surface (the Earth being a sphere of radius about 4,000 miles). Then you want the volume of a sphere of radius 4,020 miles minus the volume of a sphere of radius 4,010 miles. Since the volume of a sphere of radius r is $\frac{4}{3}\pi r^3$, this is $\frac{4}{3}\pi(4{,}020^3 - 4{,}010^3)$, or about 2 billion, cubic miles.

To do calculations such as this, in two and three dimensions, it is helpful to have handy lists of areas and volumes.

Areas:

Object	Area	Explanation
Rectangle	bh	base times height
Triangle	$\frac{1}{2}bh$	half the base times height
Trapezoid	$\frac{1}{2}(b_1 + b_2)h$	the average of the bases times height
Circle	πr^2	π times the square of the radius
Surface of a sphere	$4\pi r^2$	4π times the square of the radius

Volumes:

Object	Volume	Explanation
Cube	l^3	the cube of the side length
Rectangular box	lwh	length times width times height (or "base area" times height)
Sphere	$\frac{4}{3}\pi r^3$	$\frac{4}{3}\pi$ times the cube of the radius
Cylinder	$\pi r^2 h$	base area times height (true for "cylinder" with any base shape)
Cone	$\frac{1}{3}\pi r^2 h$	one-third base area times height (true for "cone" with any base shape)

As an example, to calculate how much water is in a pool with five-foot vertical walls and a base (of any shape) with area $600\,\text{ft}^2$, we use the cylinder formula: $600\,\text{ft}^2 \times 5\,\text{ft} = 3000$ cubic feet.

Exercises

1 You love ice cream. At the shop, you have a choice between a spherical scoop with diameter 3 in or just a filled cone leveled off flat at the top (4 in high with a diameter of 3 in at the top). Which gives you more ice cream?

2 A tennis ball is in the air at high noon. How much bigger is its surface area than its shadow?

3 My family of four wants to save money on airline baggage fees, so we carry our luggage on board for our weeklong vacation. Restrictions on the size of liquid containers means that we only pack a $100\,\text{ml} = 100\,\text{cm}^3$ container of toothpaste. Assuming we use a small cylinder of toothpaste each time we brush (you can pick a reasonable size), will we have enough?

5 Units and Scientific Notation

In the real world, numbers come from measurements. They are amounts – huge or tiny, precise or rough. We must convey all this in how we express them.

5.1 Units

We use numbers to model amounts of *things*: five apples; four pianos. You can't add these two unless you reclassify apples and pianos as part of the same class, e.g., objects: nine objects. So numbers in the real world come attached to units, and you can only combine like units. Note the similarity with algebra! For example, you wouldn't add someone's age plus their height. My age is 46 years and my height is 6 feet – so, 52? My age is also 0.46 centuries and my height is 72 in – so, 72.46? Doing this is as wrong as saying $5x + 4y = 9y$. It rightly looks like a muddle. You can't even add two of the same *kind* of quantity (such as your height and my height) unless the units are the same: feet and feet, inches and inches.

How did we go from 6 feet to 72 inches? Plainly, if each foot is 12 inches, then 6 of them make $6 \times 12 = 72$ inches. More formally, we used a *conversion*: 1 foot is 12 inches, or 1 ft = 12 in. If we continue the analogy with algebra, we can "divide both sides by 1 ft" and say 1 = 12 in/ft. Note that the number on the left has no units!? This means we can multiply anything by 12 in/ft and it will remain the same. If you multiply 6 ft by 12 in/ft, you get

$$6\,\text{ft} = 6\,\text{ft} \times 1 = 6\,\text{ft} \times \frac{12\,\text{in}}{1\,\text{ft}} = 72\,\text{in}.$$

Note that the ft unit "cancels." The number 12 here is sometimes called a "conversion factor." This can get confusing, since if we need to convert from inches to feet we multiply by the reciprocal $\frac{1\text{ft}}{12\text{in}}$ so that "inches cancel." The way to be sure is always to keep the units in your expressions. It may seem tedious, but a stitch in time saves nine. (Nine what? *Stitches!*)

Now inches (in), meters (m), and feet (ft) are different units for the same type of thing: *length*. Seconds (s) and years (yr) measure *time*. Kilograms (kg) and grams (g) measure *mass*. Conversion factors relate different units of the same type. Mass, length, and time are the three the basic types; many other units are actually combinations of these. For example, a food Calorie (Cal) is a measure of energy, equal to about $4{,}184\,\text{kg}\,\text{m}^2/\text{s}^2$. Don't worry if the combination $\text{kg}\,\text{m}^2/\text{s}^2$ looks weird. That's why people call it by another name: "joule" (J). The relation 1 Cal = 4,184 J helps us convert.

Piling up the units can make things more complicated than the above simple example suggests. Here's a tricky example. In Germany they like to measure fuel efficiency by the number of liters of gas needed to drive 100 kilometers – the lower the number, the better. In the US it is measured in miles per gallon – the higher the number the better. Suppose a car needs 8 liters to go 100 km: 8 liters per 100 km means $\frac{8L}{100km}$ (note *per* means *divided by*). How do we compare with mpg? We look up that 1 gallon = 3.73 liters and 1 mile = 1.6 kilometers. So 1 = 3.73 L/gal and 1 = 1.6 km/mi. Note that since the reciprocal of 1 is 1, we also know $1 = \frac{1}{3.73} \frac{gal}{L}$. In short, you're good if the numerator and denominator represent the same quantity. Together these numbers give

$$\frac{8}{100}\frac{L}{km} = \frac{8}{100}\frac{L}{km} \times \frac{1}{3.73}\frac{gal}{L} \times \frac{1.6}{1}\frac{km}{mi} = \frac{8 \cdot 1.6}{100 \cdot 3.73}\frac{gal}{mi},$$

where in the last step we have "cancelled" the units of km and liter. The result is about 0.0343 gal/mi. To find mpg or mi/gal, we take the reciprocal and find about 29 mpg.

5.2 Scientific Notation

In everyday life, we meet lots of big numbers: the mass of the Earth in kilograms, the number of atoms in a balloon, the number of people in India, the national debt in dollars. The mass of the Earth is about 5,972,000,000,000,000,000,000,000 kilograms. Those zeroes aren't precisely zeroes, but we can't really measure it accurately beyond the numbers listed. Two problems arise: first, that number is too big to carry around; second, those zeroes are disingenuous. To solve the first problem, we use the fact that $1,000,000,000,000,000,000,000,000 = 10^{24}$, so we can write

$$5,972,000,000,000,000,000,000,000 = 5.972 \times 10^{24}.$$

This form, called *scientific notation*, can be used to express very large or very small numbers – we simply change the exponent accordingly (for example $10^{-6} = 1/1,000,000$, one-millionth). It also solves the second problem: if we don't know the number accurately, we drop the terms that we don't know, as we have done above. Conversely, if we *did* know that the digit after the 2 was 0, we would write 5.9720×10^{24}, retaining five *significant digits* (see also Section 5.3).

As another example, at the time of writing (2015), the US national debt is *roughly* $18.6 trillion, but an exact accounting is impossible – and anyway, the exact number will change before you can even use it in a calculation! In such cases, we realize we don't want/don't need/cannot make an exact calculation, so we settle for only keeping some of the digits. In the case of the debt, $18.6 trillion can be written as 1.86×10^{13}.

We also meet numbers with more decimal places than we care to, or can, include in our calculations, for instance, $\pi = 3.141592653589\ldots$ or $e = 2.71828182845905\ldots$. We usually truncate these figures, keeping as many significant digits as we have in the most precise quantity that we shall use. In measuring the area of a circle to three significant digits, we will typically set $\pi = 3.14$. (Sometimes, because of the possibility of rounding errors, we keep a digit or two more, but only present an answer with as many significant digits as the weakest link in our calculation. See Section 5.3.)

Example A1 Here are some examples of huge, or tiny, numbers that arise in practice:

- the Gross Domestic Product (GDP), global economic figures
- Avogadro's number in chemistry, or the number of atoms in a gram of hydrogen
- the number of bytes of storage in the hard drive of a supercomputer; the number of calculations a supercomputer performs each day
- the chance of winning the Powerball jackpot
- the charge on an electron in Coulombs
- the length of a nanotube in meters
- the number of stars in the universe
- the number of cells in the liver

▲

When we talk about very large or small figures, we often speak of how many digits the number has. ("Her salary hit six digits last year!") For tiny figures such as 0.00000000437, we speak of how many places after the first nonzero digit appears. In both of these cases, it is basically the power of ten closest to the number which is relevant (maybe off by one), or put differently the logarithm of the number (see Appendix 2.5). For example, $1,000,000 = 10^6$ and $0.00001 = 10^{-5}$. Adding digits to this is fine-tuning, so for example Avogadro's number is about 6.022×10^{23}. So, scientific notation means expressing a figure in the form of a number between 1 and 10 times a power of 10. This makes performing calculations with large numbers simpler and focuses attention on only the most relevant aspects of a number: how big it is and how accurately we know it.

If we wanted to state a more precise form for Avogadro's number, for example, we would add more digits to the first factor, as in 6.022140×10^{23}. Note the zero at the end means that we have confidence in that seventh significant digit.

5.3 Messy Numbers and Significant Digits
You look out into a crowd attending a political rally and ask how many people there are. Here are three possible answers: "thousands," 5,000, 5,109. Which one is best?

Think of your response and jot it down.

For most purposes, the first answer is too cold, the last answer is too hot, and the one in the middle is just right. That is, saying "thousands" is too vague – you can't do anything with it like estimate how many buttons the candidate should bring to give away – while 5,109 is too specific: could it possibly make a difference to anyone if the answer were 5,108 instead of 5,109? Anyway, the number can change over time as people flit in and out. The middle number is probably the best. Or perhaps we require just a bit more specificity, such as 5,100 instead of 5,000 – so the first *two* digits might be even more useful.

Scientific notation is a way of keeping track of only a desired number of "significant" digits. In this example, we'd say 5,000 is accurate to one significant digit, while 5,100 is accurate to two significant digits. To highlight the level of precision, overall size of the number is indicated by a power of ten (3 here), so in the first case we write 5×10^3 instead of 5,000, and in the second we write 5.1×10^3. The exact number 5.109×10^3 would be used only if it were reliably measured and needed to that level of precision. If the people were counted individually but there was some uncertainty about just a few people in the crowd coming or going, we might use the figure 5.11×10^3. In science or any data-driven field, measurements are taken with a degree of accuracy indicated by the number of significant digits reported. If you want to convey the quantity one hundred, writing 100 is ambiguous: which of those zeroes are significant? Scientific notation removes this ambiguity. The quantity one hundred would be written $1. \times 10^2, 1.0 \times 10^2$ or 1.00×10^3, respectively, indicating one, two or three significant digits.

Here's an odd thing: if you multiply 2.4×10^2 by 3.6×10^3 the result is precisely 8.64×10^5 but should be reported as 8.6×10^5, since there would be no point in retaining the third significant digit arising from multiplying quantities that themselves were only reported with two significant digits. Odder still is addition. If I add 7.426×10^{-3} to 6.12×10^4, the result is still 6.12×10^4, since the added term is no larger than terms already being ignored! Same if I had added 1.9×10^{-6}. In general, the sum should be performed and only reported to the least precise term, in this case the hundreds represented by the 2 in 6.12×10^4. Indeed, if you are reporting that 5,000 people attended the rally and then get the news that a busload of 42 more has just arrived, the new tally is still five thousand.

In practice, if you have a series of multiplications to do and you know some of them to a higher degree of accuracy than others, do the calculation, but only report the answer to the same level of accuracy as the least-known quantity.

Exercises

1 Which conversion would you use to convert between the following units?
   a m to km
   b ft to mi
   c L to gal
   d s to wk
   e kg to lbs (on Earth)

2 Write the power of ten that corresponds to each number or prefix. For example, centi- corresponds to 10^{-2}.
   a million
   b trillion
   c milli-
   d kilo-

e micro-

f nano-

g mega-

3 A snail crawls steadily across the side of my 18-inch aquarium in 6 minutes. What's its speed in miles per hour?

Hint: One mile is 5,280 feet.

4 One Calorie is equal to 1 kilocalorie, or 1,000 calories. (Check your food label; you'll see that "Calorie" is spelled with a capital C.) These are units of energy. One calorie is 4.1868 joules. For each kilogram you lift up one meter, you expend about 9.8 joules of energy. (A kilogram weighs about 2.2 pounds.) At the gym, you do 30 deadlifts, lifting 150 pounds up one meter off the ground each time. How many total Calories of energy does that require?

5 Compute the following quantities, expressing the answer in scientific notation with the correct number of significant figures.
   a $2.12 \times 10^5 + 0.0782$
   b $0.0300 \times (2.13 \times 10^{-3})$
   c 1.234×57.89
   d $0.00003500 \times (3.211 \times 10^{-13})$.

6 My ten-year-old son is five feet tall and weighs 90 pounds. He has a normal body build. If he grows to six feet tall at the same proportions, how much will he weigh as an adult? Are you tempted to draw any conclusions from this?

Hint: If you double the size of a cube, you multiply its volume by 8, not 2. Now assume that my son is made up of tiny cubes, and you can even form one big cube from them – just put him back when you're done!

7 My friend showed me a sealed bottle of water that had been in his office for five years. It was about two ounces shy of full. Since the bottle was sealed, we will presume that the water had seeped out, somehow. Calculate how many molecules of water must have exited the container per second, on average.

8 Suppose we have a chemical reaction where 1 molecule of substance A and 2 molecules of substance B react to form 1 molecule of a product, C. Since the mass of atoms of A and B might differ, 1 gram of A will not necessarily react with 2 grams of B. Thus, in order to make calculations less cumbersome, chemists developed a unit of measure called the *mole* which keeps track of the number (rather than the mass) of basic particles (atoms, molecules, ions, etc.) in a substance. There are approximately 6.022×10^{23} (also known as Avogadro's number) particles in one mole of a substance. Therefore, just as 1 molecule of A will react with 2 molecules of B, 1 mole of A will also react

with 2 moles of B. The molarity (or molar concentration) of a solution is the number of moles of solute (the substance being dissolved) per liter of total solution. For example, if a sugar solution has 0.75 moles of sugar per liter of total solution, then we have a 0.75 molar sugar solution (denoted "0.75M"). This means that the molarity of the solution is 0.75.

Using these new definitions of moles and molarity, if the volume of our 0.75 molar sugar solution is 2.0 liters, how many total sugar molecules are in the solution?

9   270 grams of table sugar ($C_6H_{12}O_6$) are dissolved in water such that the volume of the final concentration is 6 liters. A mole of sugar has mass 180 grams. What is the molar concentration (or molarity) of the final solution?

6 Functions

In practical problems, we study systems, events, transformations, procedures, whatever: processes where something *happens*. When something happens, there is change. The state of things at the end of the process may be different from the state at the start:

$$\text{START} \rightsquigarrow \boxed{\text{something happens}} \rightsquigarrow \text{END.}$$

We might model the state of the system at the start and the end by some data. For example, the system might be your room, with the state of the system represented by the name and position of everything in it. The "something happens" might be that your younger brother comes in for a half-hour. The state of the room may be very different at the end of the process. If the state of the system is modeled by a single number at the start and end, then the process may be modeled by a *function*. For example, if we are discussing your bank account, the balance can be used to describe the initial state. The event may be the accrual of interest, and your balance afterward would be the final state of the system. In this case, if you are earning 2% on your balance, then the function which models this event is "multiply by 1.02" (see Appendix 2.4). Because I want to describe this process without knowledge of what your balance at the start might be, I need to model the input with a *variable*:

$$x \rightsquigarrow \boxed{\text{multiply by 1.02}} \rightsquigarrow 1.02x.$$

I could have described the same function using any letter: x was just a convenient choice.

Another process might be modeled by the "square it" function:

$$x \rightsquigarrow \boxed{\text{square it}} \rightsquigarrow x^2.$$

Here are some other examples of inputs and outputs which that be modeled by functions:

In	Out
age	height
weight	chance of dying of heart disease this year
length of edge of square	area of square
month	value of dollar versus yuan
atomic number	atomic mass
date	population of the US
genetic marker	chance of becoming obese
person	vote in presidential election of 1976
time in free fall in seconds	distance in meters fallen
number	five times the square of that number

So a function is a rule for producing an output number given an input number. If we call the function f and call the input number x, the output is written $f(x)$:

$$x \rightsquigarrow \boxed{\qquad f \qquad} \rightsquigarrow f(x).$$

If f is the "square it" function, we would have $f(x) = x^2$. For example, $f(3) = 3^2 = 9$. For another example, let t be the number of days since you opened your bank account, and let $B(t)$ represent the balance (in dollars) in your account after t days. Then the balance after one week is $B(7)$. The *average* closing balance in your account in the *second* week would be $\frac{1}{7}(B(8) + B(9) + \cdots + B(14))$, i.e., one-seventh of the sum of balances from the eighth day through the fourteenth.

Sometimes there are multiple inputs. You know that the area of a rectangle is base times height. Here we model a procedure by which we take the information about the shape of the rectangle – its base length b and its height h – and produce a quantity describing its overall area. We can call the function modeling the area by the letter A, so we would have $A(b,h) = bh$. (This is often written in familiar shorthand: $A = bh$.) This would be called a *function of two variables*. There are many other varieties of functions. From our discussion it should be clear that they are indispensable for modeling an enormous range of applications of quantitative reasoning.

6.1 Graphs

How do you picture a function? If we call the function f, then for each value x of the input, we get a value $f(x)$ of the output. The pair of numbers can be plotted in the plane, and all such pairs comprise the *graph of the function*. Put differently, the points of the graph are all points (x,y), where y is the output when x is the input. In other words, $y = f(x)$.

Take the "square it" function, so $f(x) = x^2$. The points where $y = f(x)$ are the solutions to the equation $y = x^2$. In other words, all points of the form (x, x^2). Here's a chart with some of them.

x	$y = x^2$
-3	9
-2	4
-1	1
0	0
1	1
2	4
3	9

We plot these points (bullets) and many more (the curve) in Figure A3.

If you wanted to know the side length of a square with area 8, you can use the graph to approximate the answer. Go up the y axis to the point where $y = 8$, and draw a horizontal line. It hits the graph in two places. Drop down from these points to find the x values. They're around ± 2.8. Only one of these is a possible length, since lengths must be positive. We have therefore approximated

$$\sqrt{8} \approx 2.8.$$

Of course, we could have used the calculator instead of making this approximation by hand, but then we wouldn't have learned anything about interpreting graphs. How good was our approximation? Cue the calculator! (Or just do it yourself.) $2.8 \times 2.8 = 7.84$. Not too bad.

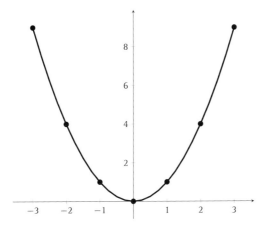

Figure A3 Graph of the function $f(x) = x^2$ for $-3 \le x \le 3$.

Presenting Graphs

Sometimes a function is defined only at select values along the x-axis. For example, if you take a reading of someone's blood oxygen levels every hour, you will have data that may look something like $\{81,90,83,82,85,88,87,92,90,94,97\}$ and can present them as a bar graph or a line graph:

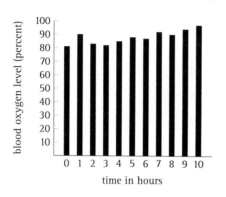
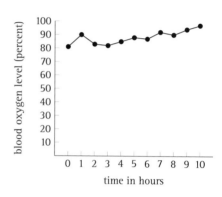

The actual data in the line graph on the right are given by the black dots.

You can understand a graph well by first looking at a *single* point of it and making sure you understand everything about it. Let's look at the datum 83 in our data set. It is not just an 83 but also the third measurement, corresponding to the third time you recorded someone's blood oxygen level. Note that this corresponds to the data point $(2,83)$ since we make the first recording at "time zero." Therefore the full data set should actually consist of points $(0,81),(1,90),(2,83),\dots.$ We also must observe that the "2" in "$(2,83)$" means two *hours* after the initial reading. Without this crucial piece of information, there is no good way to interpret the data. As well, we have to understand precisely what the 83 means. It is the percentage

Which Graph?
You have some data to present, maybe the federal budget, the price of renting a studio apartment in different cities, or the average temperature in Chicago by date. How will you display this information? A graph is a much better choice than a table of raw numbers, but which type of graph and how do you make it?

If the data are representing how a whole is divided into parts (like a budget, electoral votes, or demographics), consider a pie chart, which shows relative sizes and the fraction of the whole in a nice visual representation. Still, you probably want to include the actual numbers representing the percentage of the pie that each sector fills. Finally, be sure to add a label explaining the total amount being considered, and in which units.

If the data depend on a small number of inputs, like the example of apartment prices in different cities, then a bar graph may be the best choice. Simply list the cities on the of hemoglobin sites carrying oxygen, hemoglobin being the protein responsible for transporting oxygen in the bloodstream. (We couldn't have been talking about the percentage of *blood* which is oxygen, for then what would 100 mean!?) While not crucial to the graph, but crucial to interpreting it, the reader should know what a "normal range" is (here, 95–100; under 90 is considered low). Now that we understand the point (2,83) we also understand (5,88) and every other data point!

horizontal axis and the rent prices (in dollars per month, say – label this clearly!) along the vertical axis.

Finally, if the horizontal values are numbers on a line, you can draw a line graph or graph of all points, depending on whether the intermediate values could be suggested to be in between, as with the temperature example. As a counterpoint, one would never draw a line from the rent value in New Orleans to the one in New York, since halfway in between wouldn't represent any potential rental value in Appalachia, say.

Note that the line graph suggests the levels of oxygen in between readings, although this doesn't represent actual data that we have. Presenting the data as such can be helpful in visualizing what *might* have happened in between readings but can also be confusing to the reader, especially if fact and fiction are blurred by the presenter. Note, too, that in both graphs, having the full range of percentages from 0 to 100 is overkill, as all the action happens between 80 and 100. A compressed range can be presented, but this, too, may cause confusion if the reader sees a value of 80 near the bottom and thinks "zero." Compressed ranges can also be used deceptively, e.g., to intentionally overdramatize small effects (such as blips in sales in a quarterly report).

Looking at a graph can give you a visual sense of the average recording (here it appears to be in the high eighties) – but is no substitute for a calculated number (88.1). Looking at the graph can also quickly tell you where there was the highest rise (in the first hour) or the steepest fall (the second hour), what the trend was (increasing) and where the maximum (hour 10) or minimum (hour 0) values were recorded. We will discuss more about interpreting data in Appendix 8.

When making a graph, don't leave the reader hanging. Provide all data. Label the axes, give hash marks, make the range of numbers clear.

6.2 Linear Functions

Many quantities grow by a fixed amount in a fixed amount of time. These are modeled by linear functions.

Maybe you are a salaried worker. Then your income grows by a fixed amount of dollars in a fixed amount of time worked, say $17 per hour worked. If you start the day at time 0 with an income of $400, then your income in dollars after t hours will be 400 plus your earnings of $17t$:

$$I(t) = 400 + 17t.$$

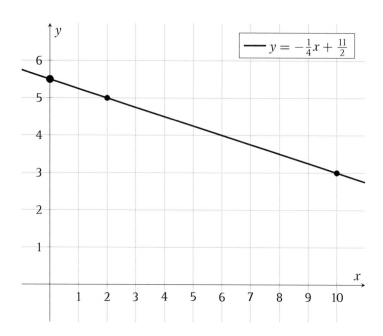

Figure A4 Graph of the linear function $f(x) = -\frac{1}{4}x + \frac{11}{2}$. The slope is $-\frac{1}{4}$ and y-intercept (large dot) is $\frac{11}{2}$. The two given points (see text) are shown (smaller dots).

More generally, if you start the day with I_0 dollars (the subscript indicates the starting time, zero) and salary rate r, your income would be $I_0 + rt$, and this gives the general form of a linear function. The rate controls how quickly your salary grows. In order not to prejudice our intended application to incomes, rates and times, we can write a linear function f as function of the variable x as

$$f(x) = mx + b.$$

The number b represents the starting value at $x = 0$, and m represents how much the function increases each step, since if you increase your input x by one unit, your output will increase by m units. (Why? Because distributivity tells us that $m(x+1)$ is m units larger than mx.) When we graph this function, it will be steeply increasing from left to right if m is large and positive, so m is called the *slope*. When x is zero the value of the function is b. Therefore the graph $y = f(x)$ hits the y-axis at the point $(0,b)$, and so b is called the *y-intercept*.

You know that two points determine a line, right? That means that given two points, such as $(2,5)$ and $(10,3)$, there is a unique linear function whose graph passes through both.[15] In this case, the y value goes from 5 to 3, a change of -2, as the x value changes by 8 from 2 to 10. We write $\Delta y = -2$ and $\Delta x = 8$. The rate of change is $-2/8 = -1/4$, and this is the slope: $m = \frac{\Delta y}{\Delta x} = -\frac{1}{4}$. We can find the y-intercept by walking x *backward* from 2. Then the y value will change from

---

[15] Disclaimer: if the points have the same x value, then the line through them will be vertical, defined by an equation $x = $ constant, thus not the graph of any function. Other than this vertical exception, though, we can find the desired function.

5 by -2 times the slope (why?), thus by $-2 \cdot (-1/4) = +\frac{1}{2}$, giving a y-intercept of $\frac{11}{2}$. So the equation is

$$y = -\frac{1}{4}x + \frac{11}{2}.$$

In general, we know that if a line passes through (x_0, y_0) with slope m, then since the rate of change is constant, all points (x, y) on the line satisfy

$$m = \frac{\Delta y}{\Delta x} = \frac{(y - y_0)}{(x - x_0)},$$

and this gives the "point-slope" form of the equation of a line: cross-multiply to find

$$y - y_0 = m(x - x_0).$$

In the example, with $(x_0, y_0) = (2, 5)$ and $m = -\frac{1}{4}$, this form of the equation would say $y - 5 = -\frac{1}{4}(x - 2)$, and a bit of algebra shows that this agrees with our previous derivation.

Remark A3 All functions are linear! Well, not really. Lots of them are curvy, like the parabola $y = x^2$ of Figure A3. However, it *is* true that nice functions look linear if you zoom into them with a microscope. Let's do that by zooming in on the parabola near the point $(-2, 4)$. Figure A5 shows how the parabola looks very much like the line $y = -4x - 4$ near this point.

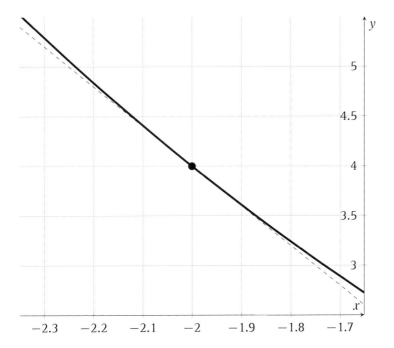

Figure A5 Graph of the function $f(x) = x^2$ for $-2.3 \le x \le -1.7$, along with an approximating line (dashed). Beware: the dashed line has slope -4, though it seems less slanted due to the different scales of the coordinate axes.

The lesson is, if we are only interested in a range of values near to some known point, we may approximate a function with a linear one. Of course, the linear function will be different if we focus on values near a different point. For example, if we look at our parabola of Figure A3 near the point $(1,1)$, the approximating line would have positive slope because the parabola is increasing there.   ▲

6.3 Quadratic Functions

Linear functions have an x term in them. Quadratic functions are the next step up: they have an x^2. To visualize them, the main thing is to recognize that the graph of x^2 is a parabola as in Figure A3. The graph of $-x^2$ is a downward pointing bowl, since the y values are reversed. The graph of $2x^2$ is a steeper bowl since all y values are doubled, while $\frac{1}{2}x^2$ is a shallower bowl. If we graph $\frac{1}{2}x^2 + 3$ the whole shallow bowl is lifted up by 3 units since all the y values are 3 higher. So much for all functions of the form $ax^2 + b$: the sign of a tells the direction the parabola points, the magnitude or absolute value of a controls how steep it is, and b shifts it up or down. All graphs of quadratic functions are parabolas. (Beware, nonquadratic functions may have bowl-shaped graphs, but they will not be parabolas.) The key is to figure out where to put the bowl, and how steep to draw it.

What about the graph $y = (x - 1)^2$? For a given x, to find the value $(x - 1)^2$, you first decrease x by one and then go up as high as the square-it function tells you. That means that if you shift this graph to the left by one unit, you'd have the graph of $y = x^2$. So this graph is the $y = x^2$ graph shifted to the *right* one unit! That rightward shift is always a bit confusing, so do a reality check: when $x = 1$, the value of $(x - 1)^2$ is zero.

To graph the function $(x - 3)^2 - 4$, we could take the parabolic graph of x^2 (see Figure A3), shift it to the *right* three units, then move it down by four. Therefore, its lowest/minimal value becomes -4, and this occurs when $x = 3$. It evidently crosses the x-axis at two points. These are where the y value is zero, i.e. when $(x - 3)^2 = 4$. This means $(x - 3) = \pm 2$, so x is either 1 or 5.

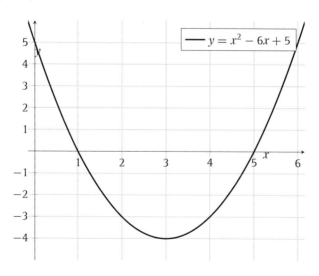

If you expand $(x-3)^2 - 4$ using FOIL for the first term, you get $x^2 - 6x + 9 - 4$, or $x^2 - 6x + 5$. This means that the graph of $x^2 - 6x + 5$ is the same as the graph of $(x-3)^2 - 4$. If you were presented with the task of graphing $x^2 - 6x + 5$, it might not be so apparent what to do unless you were able to go backwards and rewrite that expression in its more helpful form.

To do this, we'll need a few algebraic manipulations. Let's learn by example. Suppose you want to graph $x^2 + 10x - 7$. You want to write it as $(x + something)^2 + something\ else$. FOIL tells you that when you expand you get x^2, plus a number, plus a term involving x – and the number before x is *twice* what we called *something*. So we guess $(x+5)^2 + something\ else$. This gives $x^2 + 10x + 25 + something\ else$, and we find that the *something else* must be -32 to bring 25 down to -7:

$$x^2 + 10x - 7 = (x+5)^2 - 32.$$

The graph is our standard parabola (the graph of x^2) shifted to the *left* five units and *down* 32 units.

In general, finding the *something* is easy: it's half the coefficient of the x term. Then the *something else* is what fixes the constant term to work out. Let's practice with $x^2 - 14x + 50$. We want to write this as $(x-7)^2 +$??? and this is $x^2 - 14x + 49 +$???, so the additional ??? must be 1 to get that 49 up to 50. We have found $x^2 - 14x + 50 = (x-7)^2 + 1$, and the graph is our standard parabola shifted right seven units and up one unit.

Things get trickier if you have something like $3x^2 + 2x + 6$. I'd first divide everything by 3 and consider instead the simpler problem of $x^2 + \frac{2}{3}x + 2$ and go through the process – this one equals $(x + \frac{1}{3})^2 + \frac{17}{9}$ – and then multiply back by 3 to get $3(x + \frac{1}{3})^2 + \frac{17}{3}$. The parabola is steeper due to the stretching factor of 3 in front; it is shifted to the *left* by $\frac{1}{3}$, and is then moved *up* by $\frac{17}{3}$ units.

This whole process is called *completing the square*. It's useful for graphing a quadratic function and finding its minimum or maximum, which is determined by the shift. So $x^2 - 14x + 50 = (x-7)^2 + 1$ has its minimum at $x = 7$. If the coefficient of the x^2 term is positive, the graph will have a minimum; if it is negative, then the parabola gets flipped upside down and has a maximum.

A third form of our example of $x^2 - 6x + 5 = (x-3)^2 - 4$ from above is obtained by factoring. We can write this as $(x-1)(x-5)$ (check using FOIL), and this has value zero precisely at the two roots $x = 1$ and $x = 5$. We saw above that these were the zeroes, since both values satisfy $(x-3)^2 = 4$. Note that the average of these two zeros is 3, and that's because $x = 3$ is a line of symmetry for this graph. This is a general feature: the vertical line of symmetry is at the average of the roots.

What are the zeroes of a general quadratic expression $ax^2 + bx + c$? (We assume $a \neq 0$ or else it's a linear expression whose zero is easy to find.) Like I said above, I'd first divide by a and consider the easier expression $x^2 + \frac{b}{a}x + \frac{c}{a}$. Dividing by a certainly does not affect the location of the zeros, since $0/a$ is still 0. Then our rule says that the *something* is half of $\frac{b}{a}$, so we try to write

$$x^2 + \frac{b}{a}x + \frac{c}{a} = \left(x + \frac{b}{2a}\right)^2 + \text{something else.}$$

Evidently (by combining like terms), we see that the *something else* is equal to $\frac{c}{a} - (\frac{b}{2a})^2$:

$$x^2 + \frac{b}{a}x + \frac{c}{a} = \left(x + \frac{b}{2a}\right)^2 + \frac{4ac - b^2}{4a^2},$$

where in the last term we performed the square and found a common denominator. The two zeroes occur when $x + \frac{b}{2a}$ equals plus-or-minus $\frac{\sqrt{b^2-4ac}}{2a}$, where we noted the denominator was a perfect square. Solving for x and putting everything over a common denominator,

$$x = \frac{-b \pm \sqrt{b^2 - 4ac}}{2a}.$$

This is called the *quadratic formula*. The average of the two roots is $-\frac{b}{2a}$ the vertical line of symmetry of the parabola (when graphing, remember to multiply back by a). For example, the expression $x^2 - 6x + 5$ has $a = 1, b = -6, c = 5$, so the zeroes are when $x = \frac{-(-6)\pm\sqrt{(-6)^2-4\cdot1\cdot5}}{2\cdot1} = \frac{6\pm\sqrt{16}}{2} = 3 \pm 2 = 1, 5$. Indeed, we can verify with FOIL that $(x - 1)(x - 5) = x^2 - 6x + 5$, and the values 1 and 5 are where one of the two factors equals zero.

6.4 The Exponential Function

Many quantities grow or shrink by a fixed *percentage* (or fraction, or factor) in a fixed amount of time. These are modeled by exponential functions.

That simple statement is very powerful and captures it all, though it will take a little time and effort to appreciate. If a population is growing exponentially and doubles in a decade, then in the next decade it will double again. That means it quadruples in twenty years. In 30 years, it becomes *eight* times as large (not six!), as it will have doubled, doubled, and doubled.

What will it do in 5 years' time? Well, it must grow by *some* factor, since that's the property of exponentials. We don't know what that factor is, so we'll use a variable x until we can determine what x is. If we start with a population of 100 and it grows by a factor of x every 5 years, then after 5 years, it will be at $100x$, and after another 5, it will be at $(100x)x = 100x^2$. But we already know that our population must have doubled after those ten years, so $100x^2 = 100 \times 2$, or canceling the 100, we get $x^2 = 2$. So $x = \sqrt{2} \approx 1.41$, and so the population grows by a factor of 1.41, or equivalently increases by about 41%, every 5 years. This is a little counterintuitive: the population grew by 41% in 5 years but *not* by twice that (82%) in ten years, rather by 100%. Why not? Because in the subsequent 5 years, the new 41% will be calculated based on a larger population. This observation is also the mathematical underpinning behind the exponential behavior of loan balances (see Chapter 3): interest upon interest upon interest.

Let's now observe that the number 100 played no role in the analysis: if our population grows by a factor of x in 5 years, then it will grow by another factor of x, or x^2 total, in ten years, and therefore $x^2 = 2$. So the same statement is true *whatever* the population, and *whenever* we run the argument.

Now what if we consider an arbitrary time t, not necessarily 5 or 10. Suppose we start an exponential process with a "population" of a and after every unit of time (in some units) the population doubles (so above, the unit would be decades). To write a formula for the population $P(t)$ as a function of time t, we can say the "start" is at $t = 0$, so the population goes from some value, say a, at time 0, to value $2 \times a$ at time 1, to $2 \times 2a$ at time 2, to $2 \times 2 \times 2a$ at time 3, and it should be clear that at time t we have $P(t) = a2^t$. Note that this agrees with what we found above: if we put $t = \frac{1}{2}$, then we find $P(\frac{1}{2}) = a2^{\frac{1}{2}} = a\sqrt{2}$, and the population increases by a factor of $\sqrt{2} \approx 1.41$ in half the doubling time. Above, the doubling time was a decade, so half that was 5 years.

In ten doublings, a value will increase by a factor of $2^{10} = 1024$, more than a thousand-fold! Exponentially increasing functions – even if they start out rather modest – eventually get large as their graphs grow steep!

Now if the population had multiplied by a factor of b instead of 2, then the behavior would be

$$\boxed{P(t) = ab^t.}$$

Every exponential function has this form, but you won't always see it presented as such. The reason is that for exponents and logarithms, people don't always want to mess around with lots of different bases like 2 or b or whatever. In real life, two particular bases seem to have won out: base e and base 10. Using the laws of logarithms, we can write b^t as e to *some* power, which will depend on t; we just need to figure out which. So, how do we write b^t using base e? Well, b is e to *some* power – so if we have $b = e^k$ for some number k, then $b^t = (e^k)^t = e^{kt}$. So every exponential function has the form

$$\boxed{P(t) = ae^{kt}.}$$

Note that the statement $b = e^k$ is the same as $k = \ln b$. Likewise, we could write

$$\boxed{P(t) = a \cdot 10^{ct},}$$

where now $c = \log b$ (why?). We also therefore have

$$k = \ln b = \frac{\ln b}{\log b}\log b = \frac{\ln b}{\log b}c = c\ln 10.$$

Let's check that $\frac{\ln b}{\log b} = \ln 10$ as claimed. This means (cross-multiplying) $\ln b = \ln 10 \cdot \log b$. Raising e to both sides gives $b = e^{\ln 10 \cdot \log b} = 10^{\log b} = b$, as claimed.

Exponential *decay* is when the growth factor b is less than 1, so the population is actually shrinking. This is equivalent to the coefficients k and c being negative, as logarithms of numbers less than one are negative. For example, some of

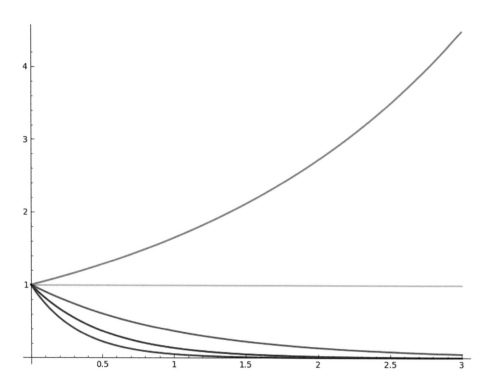

Figure A6 Graphs of the exponential function $f(x) = e^{ax}$ with a taking various values: $a = 1/2$ (top, increasing), $a = 0$ (flat), $a = -1, a = -2, a = -3$ (bottom, decreasing most rapidly). Note that for $a = -3$, the "width" of the bottom graph is roughly $1/3$, or $1/|a|$. When a is negative, $1/|a|$ is a reliable measure of its width, while the decay rate $|a|$ indicates how quickly the graph decreases.

the atoms in radioactive materials are radioactive isotopes. These isotopes decay, emitting radiation, so the material becomes less radioactive in the process. After some amount of time, the number of radioactive isotopes will become half of what it was. That amount of time is called the *half-life*. For example, in some nuclear accidents such as Chernobyl, radioactive cesium-137 is released, and this material has a half-life of 30 years. Let's do a practice problem: *assuming* that the presence of cesium is the only risk factor and that the decay rate is as expected,[16] how much safer will a nuclear disaster site be after a century?

To answer, first we must determine the population of cesium-137 as a function of time. Since we are only making a relative comparison rather than an absolute measurement of the amount of cesium-137, the initial value could be anything – but so as not to presuppose any value, we use a letter C for this value. Then if $P(t)$ is the amount of cesium-137 at time t, then at time 0, we have $P(0) = C$. We also know $P(30) = \frac{1}{2} \times C$, if we measure time in years.[17] Likewise, $P(2 \times 30) = \frac{1}{2} \times \frac{1}{2} \times C$.

---

[16] Other environmental factors are involved, so this assumption is not always valid.

[17] It might be helpful to think of time first as being measured in denominations of 30 years, then $P(T) = C \cdot (\frac{1}{2})^T$. Then our time t in the text can be thought of as $t = 30T$, or $T = t/30$.

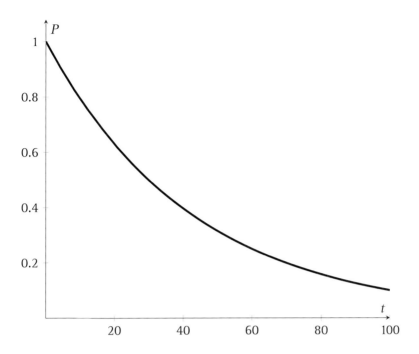

Figure A7 Graph of $P(t) = 1 \cdot (\frac{1}{2})^{t/30}$. Note the value at $t = 100$ is about 0.1.

Continuing, we see $P(30N) = C(\frac{1}{2})^N$. Now put $t = 30N$, or equivalently, $N = t/30$. This says $P(t) = C(\frac{1}{2})^{t/30}$. But $\frac{1}{2} = 2^{-1}$, so we can write this as $P(t) = C \cdot 2^{-t/30}$. If we want to write this in terms of e, we write $2 = e^{\ln(2)}$, and then

$$P(t) = Ce^{-t \ln(2)/30} \approx Ce^{-0.0231t}.$$

Back to the question. How much safer is the situation at $t = 100$? We want to compare $P(0) = C$ and $P(100) \approx Ce^{-2.31}$. This means $P(0)/P(100) \approx e^{2.31} \approx 10.1$. So after 100 years, the level of radioactivity is about a tenth of what it was initially. That is, the situation is about ten times safer after a century. We can approach the same question visually if we graph $P(t)$. In Figure A7, we use $C = 1$, because the precise value of C plays no role.

As t gets larger and larger, the value tends toward zero. Exponential increase (rather than decay) tends toward infinity over time. If you are modeling a real-life process by exponential increase, this means the model will become inapplicable at some point. For instance, if we are describing a population of fruit flies, if they truly become so numerous they would cover the earth, eventually dying off when they had nothing left to eat.

6.5 Inverse Functions, Logarithms, and More on Graphing
What positive number squares to 9? To get the answer (3), we need to run the squaring process in reverse. We use the *square-it* function, but we begin with the desired output (9) and find the input (3). It's like mathematical *Jeopardy!* and is

very useful. The process which produces the input, given the output of a function, is called the *inverse function*. (Warning: the inverse function is *not* the reciprocal of the function.)

For example, if after receiving 2% interest on your bank account, you had $459 in your account, how would you find out what you had before the interest was earned? We know from the start of this appendix that the function modeling the addition of 2% interest is "multiply by 1.02." Running this machine backwards means dividing by 1.02. Cue the calculator! $459/1.02 = $450.

A population is described by the formula $P(t) = 50 \times 2^t$. When will it hit 1000? This is an inverse problem since we want to run the P machine backwards and find out what value of t produced an answer of 1000. That is, we need to solve $P(t) = 1000$ for t, so $50 \times 2^t = 1000$, or $2^t = 20$. Now what? We need the inverse function of exponentiation with base 2. This is the logarithm, base 2. Taking $\log_2$ of both sides gives $t = \log_2(20) = \log(20)/\log(2) = 4.32$.

Inputs and outputs are reversed for the function and its inverse; accordingly, the graph of the inverse function is the same but with x- and y-axes swapped. Did you know it was so easy?

The inverse function of exponentiation by 10 is the logarithm. This means that $\log y$ is that number which, when you exponentiate it, yields y. Or $10^{\log y} = y$. Generally, if f denotes a function, then the inverse function is denoted f^{-1}. Warning: this is *not* the same as $1/f$![18] As for exponentials and logarithms, we have $f(f^{-1}(y)) = y$ and $f^{-1}(f(x)) = x$. That's because when you run a process backward then forward – or forward then backward – you return to the same place.

Returning to the exponential, since 10^x increases very quickly as x grows, the inverse function $\log(y)$ increases very slowly as y grows: indeed $\log(1{,}000) = 3$ but $\log(1{,}000{,}000)$ is only 6. Therefore, the logarithmic function is very useful for describing inputs which range over vast amounts. The following quantities are measured with logarithms or plotted with logarithmic scales: intensity of earthquakes (the Richter scale), sound intensity (decibels), evolutionary changes (plotted over time), mass of animal species (kilograms). Because the changes can be so small for small changes in input, graphs of logarithmic functions are often presented on a "log scale," where each tick measures an increase of the input coordinate (x) by a fixed *factor*, rather than a fixed value. So, successive ticks on a log scale would be values of x such as $0.01, .1, 1, 10, 100, \ldots$ since $\log(x) = -2, -1, 0, 1, 2, \ldots$. With this trick, the graph of a logarithmic function will actually look like a line. Consider, if $y = 2 + 3\log(x)$ then y is 2 when $\log(x)$ is zero, and each time $\log(x)$ increases by 1 (i.e., each "tick"), y increases by 3.

---

[18] On behalf of mathematicians all over, I apologize for the misleading notation. We're usually better than that!

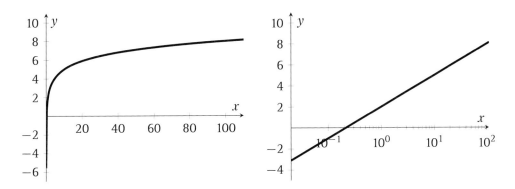

Figure A8 Here we can see the same function, $y = 2 + 3\log(x)$, plotted without (left) and with (right) a log scale for the x coordinate. Beware: the y axis at right is drawn near $x = 0.02$, not $x = 0$, which would be an infinite distance to the left in this log scale graph.

Exercises

1 What is the value of $\log_5(25)$?
   a $\frac{1}{2}$
   b 2
   c 5
   d 9,765,625

2 What is the value of $x^2 - 3x + 4$ when $x = 3$?
   a -4
   b 4
   c 7
   d 21

3 For the following functions, describe what happens to the value of the function $y = f(x)$ as x approaches ∞ and $-\infty$. Do you notice a pattern?
   a $y = x + 4$
   b $y = x^2 + 5x - 3$
   c $y = x^3 - 6x^2 + 4x + 3$
   d $y = x^4 + 6x^3 - 3x^2 + 4x - 2$

4 My bank account starts at $8,860 and decreases at a (continuous) steady rate of $105 per day. Write a formula for my balance (in dollars) as a function of time in days. Write a formula for my balance in dollars as a function of time in hours. My daughter's piggy bank account starts at $4 and increases at a continuous steady rate of $3 per week. Write a formula for my daughter's balance in dollars as a function of time in whatever units you like – but do specify. When will my daughter and I have equal balances?

5 What's the slope of the line segment connecting the points $(4,5)$ and $(10,7)$?

6 Define a variable representing the number of days since January 1, 2019. Define a function of your variable representing the cumulative snowfall in Chicago for the year 2019 as a function of the date. In term of these definitions, write an expression for the average daily snowfall in Chicago in March, 2019.

7 You run a company that makes malted milk powder for sale online. Each unit comes in a jar packed in a $6'' \times 6'' \times 6''$ box. How many cubic feet is that? Your contract with the on-line retailer requires you to deliver your product to their warehouse. To do that, you use a shipping company that charges $200 for a truck that can carry 1,000 cubic feet of merchandise, or $600 for a truck with a 5,000 cubic-foot capacity. (The price covers gas, salaries, everything.)

How many units will fit in the smaller truck? The bigger?

For this exercise, you must figure out which delivery method to use depending on how many units you are shipping. You will also need to determine the per-unit shipping cost. For example, if you are shipping one hundred units, you would hire the smaller truck for $200, and your per-unit shipping cost is $200/100 = 2.

Find a method to determine how many of which ships to use based on the number of units N that you must ship. Next, determine how many of which trucks to rent if you have 50,000 units to ship and find the per-unit shipping cost. Do the same with 100,000 units.

8 a Translate the statement *the output of the average of the inputs is the average of the outputs* into a formula for a function $f(x)$.

b Suppose $f(x)$ is a linear function. Prove that $f(x)$ satisfies your equation from Part 8a

9 The number N of radioactive isotopes in a sample of C_{11} is given by $N(t) = 10^{30}e^{-0.0341t}$, where t is the time in minutes. What is the half-life of the sample? That is, after how many minutes will the number of radioactive isotopes in the sample decrease by one-half?

10 Newton's law of cooling predicts that the temperature of an object as a function of time follows a kind of exponential decay. More precisely, we only study the temperature *above* room temperature; that is, the difference between room temperature and the temperature of the object. Common sense says that the temperature will eventually approach room temperature, so the *difference* between temperature and room temperature will eventually approach zero: good. It is this quantity (the difference) which follows an exponential decay. Now suppose room temperature is 70°F and we take a 340°F baked potato out of the oven (so its temperature above room temperature is 270°F). Ten minutes later, we measure the temperature of the potato to be 250°F. What will the potato's temperature be in another ten minutes' time?

Hint: we know its temperature-above-room-temperature cools by a fixed factor in a fixed period of time.

11 You buy a data roaming plan from your cellphone carrier. The charge is $30 for the first 200 megabytes and $5 for each additional 10 megabytes thereafter. Graph the charge as a function of the amount of data used, between 0 and 300 megabytes. At what point is it cheaper to buy an additional plan than to go with one plan with an overage charge?

12 Your hair, currently $2\frac{1}{2}$ inches long, grows at a rate of a quarter inch per fortnight. Write down your hair length as a function of time in weeks. Now suppose you cut an inch off every month (four weeks). Draw the graph of the hair-length function in the range from 0 to 20 weeks. Comments?

13 You kick off a football (from ground level), and the height of the football is given by $h(x) = -\frac{1}{50}x^2 + 2x - 32$, where x is the position, in yards, on the field. The football field extends from $x = 0$ yards to $x = 100$ yards, and you kicked the ball toward your opponent's end zone at $x = 100$. Where did you kick from? Did you make it into the opposing end zone? How high did the football go?

7 Probability

You probably think that if you toss a coin the chances are 50% that it will turn up heads. We might denote this $P(H) = 1/2$, the P for "probability" and the H for "heads." If it's a fair coin, that is the correct chance of heads, but now try this: toss it ten times and count how many times it came up heads. Do this now.

. . .

Waiting, while you do the experiment. Record your results!

. . .

I bet that you did **not** get five heads. Maybe I'm wrong, but chances are that I'm right! Why?

First let's review why we thought the probability of heads is $1/2$, aka 0.5 or 50%. There are two possible outcomes, one of which is heads. One out of two is $1/2$. Of course, when you cross the street there are two possible outcomes: you get hit by a car or you do not. And yet the probability is not $1/2$, or else you wouldn't have gotten this far in life. The point is that with the coin toss, both outcomes were equally likely.

If we roll a die and I ask, "What's the chance that I get a 3?" this is one out of six equally likely outcomes, so the chance is 1 in 6, or $1/6$: we can write $P(3) = 1/6$. And if I ask what's the chance that I roll a perfect square? Of the six equally likely outcomes, exactly two of them ($1^2 = 1$ and $2^2 = 4$) are perfect squares – so the chance is 2 in 6, or 1/3. Note that while I will either roll a perfect square or not, the answer is **not** one in two.

Summarizing: if all outcomes are equally likely, the probability of an event is the number of outcomes in which the event occurs (such as rolling a perfect square) divided by the total number of outcomes.

But if the chance of heads is one in two, why did I bet *against* getting five heads out of ten flips? To follow our rules, we should write down all the possibilities of what could happen in ten flips. This sounds hard, so let's begin with two flips. (We won't consider one flip because we could never have gotten half a head.) With two flips, we could have HH, HT, TH, or TT. Of these four equally likely outcomes, exactly two of them have one head, so the chance of getting one head in two flips is 2 in 4, or 1/2. Note this a 50-50 chance, not a guarantee. With four tosses we have

HHHH, HHHT, HHTH, <u>HHTT</u>, HTHH, <u>HTHT</u>, <u>HTTH</u>, HTTT

THHH, <u>THHT</u>, <u>THTH</u>, THTT, <u>TTHH</u>, TTHT, TTTH, TTTT

Note 6 in 16 (underlined), or 3/8 of them have exactly two heads, so already the chances are unlikely that we'll get exactly half the tosses coming up heads. While it's true that of the possible outcomes, half of them heads is the most likely single one – yet its likelihood is less than half. If we continue this analysis, we'll find $2^{10} = 1,024$ possible outcomes for a sequence of ten tosses, of which 252 had exactly five heads, meaning it will happen about 24.6% of the time.

So what does it even mean that the probability is 50%? You may have heard things like, "if you toss the coin a billion times, then half of them ..." but we already saw that it was unlikely that exactly half of them would be heads after just four tosses! To understand better, we should consider the likelihood of *other* fractions of the total number of coin tosses being heads. All such fractions lie between 0 and 1. Then we can plot the likelihoods for different numbers of total coin tosses and see what happens when we toss the coin zillions of times. A probability of one-half means that the likelihood of all other fractions besides one-half will head toward zero.[19]

7.1 Expected Value

Would you place a $10 bet on a one-in-a-hundred chance of winning $20? Most people wouldn't. Suppose you did the bet 100 times. On about one of them you would expect to win $10 (your winnings of $20 minus your outlay of $10), while you would lose your $10 about 99 times, putting you in a net deficit. In fact, we can calculate your expected winnings minus losses to be $10 - 990 = -980$ dollars. This is in 100 bets. So for a single bet, we can say you would "expect" to lose $9.80. Tracing through our calculation, we found the expected gain to be

---

[19] See, e.g., Figure A9 in Appendix 8.5, which has more on the subject.

$$\frac{1}{100} \times 10 + \frac{99}{100} \times (-10) = -9.80$$

dollars, here a negative quantity (so a loss). There were two possibilities: win or lose, and for each one we multiplied the probability times the expected gain (or loss) from that outcome.

> The expected value is the sum over all outcomes of their values times their probabilities.

Example A2 A neighbor is selling tickets for a raffle; 100 tickets will be sold for $5 each. First prize is $50, two second prizes win $25, and three third prizes win $10. What is your expected gain or loss? To answer, we need the values and probabilities of each outcome. There is a 1/100 chance of winning first prize for a net gain of $45, a 2/100 chance of gaining $20, and a 3/100 chance of gaining $5, with a 94/100 chance of losing $5. We compute the average gain to be

$$\frac{1}{100} \times \$45 + \frac{2}{100} \times \$20 + \frac{3}{100} \times \$5 + \frac{94}{100} \times (-\$5) = \frac{\$45 + \$40 + \$15 - \$470}{100}$$

$$= -\$3.70,$$

so buying a ticket is akin to making a donation of $3.70 to the cause.   ▲

7.2 Conditional Probability; Bayes's Theorem

Question: Xenia has rolled a die. What is the chance she got a 5? Answer: 1/6.

Question: Yulia has rolled a die and got higher than a 3. What is the chance she got a 5? Answer: 1/3.

What changed between the first and second question is we *knew something* about Yulia's outcome, so the range of possible values had changed. Knowing what we knew about Yulia's roll, 5 was in fact one of only three equally likely values (4,5,6) rather than one of six. This phenomenon is called *conditional probability*: it is the probability of A occurring *given that B* has occurred, denoted $P(A|B)$.

What are the chances that you roll an even number greater than 3? There are two such numbers, so the probability is 2/6 = 1/3. But we can think of this another way. You have rolled greater than 3 *and* you have rolled an even number on top of that. Now let's take these two events in turn. The chance that you rolled greater than 3 is 3/6 = 1/2. And what are the chances of rolling an even number *given that* you have rolled greater than 3? The answer is 2/3. The probability of doing both as such is (1/2)(2/3) = 1/3. Let's see how this works. If we write "> 3" to indicate the event of rolling more than 3, then we have

$$P(> 3 \text{ and even}) = \frac{\#\{> 3 \text{ and even}\}}{\text{total } \#} = \frac{\#\{> 3\}}{\text{total } \#} \cdot \frac{\#\{> 3 \text{ and even}\}}{\#\{> 3\}}.$$

More generally, we can find the conditional probability by

$$P(A \text{ and } B) = P(B) \cdot P(A|B),$$

or in other words, $P(A|B) = P(A \text{ and } B)/P(B)$.

The formula can be turned around in a useful way. We know $P(A \text{ and } B) = P(B \text{ and } A)$ since "A and B" means the same thing as "B and A." Using this simple fact with the formula above, now with reversed roles for A and B, we get

$$P(A|B)P(B) = P(B|A)P(A),$$

and upon dividing by $P(B)$, we arrive at Bayes's Theorem:

$$P(A|B) = \frac{P(B|A)P(A)}{P(B)}.$$

This is handy for when finding $P(A|B)$ is hard but $P(B|A)$ is easy – see Exercise 11.

Exercises

1 Thirty people come to a party. Two will win the centerpiece raffle, and one will be chosen to be the king/queen of the party. What is the probability of winning a centerpiece?

  a $\frac{1}{30}$

  b $\frac{2}{30}$

  c $\frac{3}{30}$

  d $\frac{4}{30}$

2 What is the probability of rolling a three on a ...

  a 4-sided die?

  b 6-sided die?

  c 12-sided die?

3 You roll a pair of fair four-sided dice. Each die can show the number 1, 2, 3, or 4, all equally likely. What's the probability that the total is greater than 5?

4 I flip a pair of (fair) coins twice. What's the chance I get the same result both times?

  Hint: the pair are indistinguishable, and I flip them at the same time.

5 I roll a pair of (fair) dice twice. What's the chance I get the same roll both times?

  Hint: the pair are indistinguishable, and I roll them at the same time.

6 There are three students in my office. None has a birthday on Leap Day, February 29. What are the chances that no two of them share a birthday? What are the chances that two *or more* of them share a birthday? Can you repeat the exercise with four students? N?

7 This classic problem is the so-called Monty Hall problem.[20] It goes like this. You are on TV dressed as Little Bo Peep. You have the choice to pick one

---

[20] Monty Hall was the longtime host of a TV game show called *Let's Make a Deal*. It is currently (as of 2019) hosted by improv comedy guru Wayne Brady.

of three doors. Behind one of them is a prize – a new car – while there is a booby prize behind two (a goat, usually – but who doesn't like goats?). Anyhoo, you want the car. You select a door. Then Monty (or Wayne) shows you a goat behind one of the doors that you did *not* select, and gives you the opportunity to switch your guess to the remaining undisclosed door. Should you do it? Does it even matter?

Hint: your costume might matter for the TV show, but not for this question!

8 You have an extra $100 ticket to the opera. You don't need it, and (for the sake of argument) you have no moral compunction about selling it illegally on the street. If you are caught selling it, you will be fined $500 and receive no money. You estimate the chance of getting caught to be about 10%. You estimate that it will take about ten minutes to sell your ticket, and for your time and effort, you want to earn at least $30 expected profit (beyond the $100 you paid). Others are selling their tickets on the street, too, so you want to choose the price that returns you an expected profit of exactly $30, no more. For what price should you sell your ticket?

9 You go to game night at the senior center, and one of the tables features a game that uses a single six-sided die. The price to play is half a dollar. If you roll a 1 or 2, game over. If you roll a 3, 4, or 5, then you get your money back. If you roll a 6, then you get a dollar back.

   a If you play this game many times, what would be your expected gain/loss per game?
   b For what price to play would you expect to break even?
   c Suppose the rules of the game are altered such that you are allowed to pick the amount x you pay to play, measured in dollars. Rolling a 3, 4, or 5 returns x, while rolling a 6 gives you back $2x$ – so the game is the same if you pick $x = \frac{1}{2}$. What is your expected gain/loss per game as a function of x? Verify that your result agrees with part (a) when $x = \frac{1}{2}$.
   d Now consider another variation where again you are allowed to pick the amount x you pay to play. Let n be the number you roll. If you roll $n = 1$ or 2, you (lose your initial payment) and must pay an additional n^2 dollars. If you roll a 3, you must pay out an additional dollar. If you roll $n = 4$, 5, or 6, you get back $n - 2$ times the price you paid. For which prices to play would you expect to make money?

10 You are approaching a green light that has just turned yellow, 130 ft ahead. In 2 seconds, the light will go red. You are traveling in a 30 mph speed zone. For the purposes of this question, let's assume there is no one around for miles and you have no compunctions about breaking the law. However, there is a traffic camera at the intersection. Based on your experiences, the probability that the camera will issue you a ticket *for speeding* is $(s - 30)/20$, where s is your speed in miles per hour between 30 and 50. For other ranges of speed, you know that below 30 you will definitely not get a ticket and above 50

you definitely will. Also for the purposes of this question we will assume that you have a digital speedometer, so that your speed will be registered only in integer values. There is another issue at play: whether you will get a ticket for running the red light. Based on your experience, the chances of getting a ticket for running the red light is $t/2$, where t is the time in seconds between 0 and 2 that the light has been red when you pass. (Above $t = 2$, the probability is 100% that you will get a ticket.) In short, the faster you go and the longer the light has been red, the higher the chance you'll get nabbed.

  a What is the chance that you will get a speeding ticket if you go 43 mph?

  b What is the chance that you will get a ticket for running the red light if you pass through the intersection 1.2 seconds after the light turns red?

  c Below what speed will you fail to make the light?

     Hint: One mile is 5,280 ft. One hour is 3,600 s.

  d You have decided to drive at 36 mph:

    (i) What is the chance of *not* getting a speeding ticket at this speed?

    (ii) What is the chance of *not* getting a ticket for running the red light at this speed?

    (iii) Assume that the two events above are *independent*, meaning one does not affect the other. What is the chance of getting off without a ticket at this speed?

11 A class of 20 boys and 30 girls are selling cookies to raise money for a field trip. Each child sells *either* chocolate-chip *or* oatmeal-raisin cookies, *not* both. 8 boys and 24 girls sell chocolate-chip cookies.

  a if a girl from this class knocks on your door, what is the chance she is selling oatmeal-raisin cookies?

  b If a child from this class comes to your door to sell chocolate-chip cookies, what is the chance it is a boy?

  c Suppose you know only that 40% of the students are boys, 64% of the sellers sell chocolate chip, and the answers to the questions above. Use Bayes's Theorem to find out the chance that if a boy knocks on your door to sell cookies, they are chocolate chip. Now square this with a computation using full knowledge of the numbers.

8 Statistics

8.1 Mean, Norm, or Average

You hear about a town where the average annual income is a million dollars. Should you move there? Upon closer inspection it turns out that of the working population of 1,001, there are 1,000 people with income of $25,000/yr and one super-rich magnate with income of $976,000,000. The average, or *mean* or *norm*, of this distribution of incomes is the sum of the incomes divided by the number of people. Adding up a thousand times $25,000 plus $976,000,000 and dividing by the population of 1,001, we get ($25,000,000 + $976,000,000)/1,001 = $1,000,000, a million dollars.

While technically true that the average income is a million dollars, the number can be misleading, as we typically think of the English word "average" as being associated to the "typical experience." Here it is very clear that the typical experience is nothing like the average, and in fact most people would be struggling: this income would put a family of four in 2017 right about at the poverty line.

The lesson is that average is *a* measure, but just *one* measure, of a distribution of numbers.

> Given a distribution of values denoted by x, the mean, denoted $\bar{x}$ or $\langle x \rangle$, is the average, i.e., the sum of all the values of x divided by the number of points in the distribution.

8.2 Median

One way around misleading means is to use the *median* value of a data set. The median is the number that is "in the middle" of the data set, meaning half the points are above and half the points are below the median value.

For example, to find the mean of $\{1,7,3,6,2\}$, we write the data in order 1, 2, 3, 6, 7, and 3 is the median. If the data are $\{5,8,6,0\}$, there is no single number, and we can define the median to be the average of the two numbers (5 and 6) in the middle: 5.5. (A more realistic data set would have lots of numbers in close proximity, so how you handle the "middle" won't matter much.) As another example, for the data set $\{8,5,7,9,7,7,1\}$, the median is 7.

For a town with many people making a meager living with a small salary and one gazillionaire, the average salary might be high, but the median would be lower and give a better picture of the "typical" earner.

8.3 Variance, Standard Deviation

Suppose there are just two students in my class. On the first test they both score 80, while on the second, one scores 60 and the other 100. In both cases, the mean is 80, but in the second test, the distribution of grades shows significant deviation from the mean or norm. We would want to assign a quantitative measure of the "spread" of a distribution. It will be called *standard deviation.*

Since standard deviation will measure the spread away from the norm, it is convenient to set the norm itself to zero by subtracting it from all quantities. In this test example, we subtract 80 from all values of both distributions, leaving us with the following two distributions of the two test scores (as measured from the mean of 80): one is 0 and 0, and the other is -20 and 20. One could say that in the first case the spread is zero, while in the second the spread is 20, so a reasonable definition of standard deviation – whatever it is – should match these values.

If x is a number, then its distance from zero is $|x|$, so if x is $+3$ or -3, we get $|x| = 3$ in both cases. Now note that $|x| = \sqrt{x^2}$. It may seem natural just to take the standard deviation of a distribution whose values away from the mean we call x to be the average value of $|x|$, but in fact we take the average value of x^2 and *then* take the square root. This actually becomes a simpler quantity to calculate, a main reason for the definition. In our examples we have the average of 0^2 and 0^2, and this is 0, which gives 0 again on taking the square root. Or we take the average of 20^2 and $(-20)^2$, and this gives the average of 400 and 400, which is 400 – so we get a standard deviation of $\sqrt{400} = 20$, as demanded. By the way, the average of x^2 is called the *variance*, so the standard deviation is the square root of the variance.

> Given a distribution of values denoted by x, the variance is the mean of $(x - \bar{x})^2$. The standard deviation is the square root of the variance and measures the spread of the distribution from its norm. Denote the standard deviation by σ. Then $\sigma = \sqrt{\langle (x - \bar{x})^2 \rangle}$.

With the data set $\{5,2,4,6,4,4,3\}$, we compute the mean to be $\bar{x} = 4$. So the respective differences from the mean are $\{1, -2, 0, 2, 0, 0, -1\}$. Their squares are $\{1^2, (-2)^2, 0^2, 2^2, 0^2, 0^2, (-1)^2\}$. These add up to 10, which means a variance of $10/7$. The square root of this is the standard deviation, $\sqrt{10/7} \approx 1.195$.

8.4 Probability Distributions versus Statistical Distributions

We have encountered two situations where a quantity like a test score or winnings can take on different values. In a probability distribution, the values occur with different probabilities, and we can calculate an expected value by taking a weighted average according to the probabilities. In a statistical distribution, we take a simple average, but certain values may occur more than others, giving a similar effect. In fact, the same quantities – mean (or norm) variance and standard deviation – can be calculated for either probability distributions or statistical distributions.

Let's do this for dice rolling. The probabilities of the six values are all $1/6$, giving an expected value (or mean) of $\frac{1}{6}(1 + 2 + 3 + 4 + 5 + 6) = 3.5$.

But if we actually roll a die six times – wait, let me do this ... – we might get a data set of $1,2,3,3,4,5$. Here we measure an average value of $\frac{1}{6}(1 + 2 + 3 + 3 + 4 + 5) = 3$.

Likewise, we can measure the variance of our probability distribution (remember, you have to subtract off the norm) to be

$$\frac{1}{6}((1 - 3.5)^2 + (2 - 3.5)^2 + (3 - 3.5)^2 + (4 - 3.5)^2 + (5 - 3.5)^2 + (6 - 3.5)^2)$$

$$= \frac{35}{12} \approx 2.9,$$

and the standard deviation is $\sqrt{35/12} \approx 1.7$.

The variance of our data set is $\frac{1}{6}((1 - 3)^2 + (2 - 3)^2 + (3 - 3)^2 + (3 - 3)^2 + (4 - 3)^2 + (5 - 3)^2) = \frac{10}{6} \approx 1.7$. and the standard deviation is then $\sqrt{10/6} \approx 1.3$. Eyeballing our data set, it seems more clustered around 3 than the probabilities are around their mean of 3.5, so it makes sense that the standard deviation is smaller.

In practice, you might be polling 450 people about their salary and plot the answers on some bar graph. This would be your data set, and you can derive from it the average salary and the standard deviation. These are two measures that might give you a *feel* for what the graph might look like, but they are not a true substitute for the actual data.

8.5 Normal Distribution; Central Limit Theorem

If we're playing Monopoly, we roll *two* dice and take their sum. The totals range from 2 to 12 (underlined here for clarity), and occur with the following probabilities:[21]

$$\left\{ 2 : \frac{1}{36}, 3 : \frac{2}{36}, 4 : \frac{3}{36}, 5 : \frac{4}{36}, 6 : \frac{5}{36}, 7 : \frac{6}{36}, 8 : \frac{5}{36}, \right.$$
$$\left. 9 : \frac{4}{36}, 10 : \frac{3}{36}, 11 : \frac{2}{36}, 12 : \frac{1}{36} \right\}.$$

The norm of this probability distribution is the sum over values times their probabilities:

$$\frac{2 \cdot 1}{36} + \frac{3 \cdot 2}{36} + \frac{4 \cdot 3}{36} + \frac{5 \cdot 4}{36} + \frac{6 \cdot 5}{36} + \frac{7 \cdot 6}{36}$$
$$+ \frac{8 \cdot 5}{36} + \frac{9 \cdot 4}{36} + \frac{10 \cdot 3}{36} + \frac{11 \cdot 2}{36} + \frac{12 \cdot 1}{36} = 7.$$

This is a general feature: the norm of the sum (here the sum of two rolls) is the sum of the norms. Here $3.5 + 3.5 = 7$.

Let's calculate the variance. First we subtract 7 from each quantity, then square the result. This is multiplied by its probability, and we add to the sum, so the first term is $(2-7)^2 \times \frac{1}{36}$, and then we add $(3-7)^2 \times \frac{2}{36}$, and so on. The result is $\frac{1}{36}(25 + 32 + 27 + 16 + 5 + 0 + 5 + 16 + 27 + 32 + 25) = \frac{210}{36} = \frac{70}{12}$, and we see that the variance of the sum of two rolls $x_1 + x_2$ is twice the variance of a single roll of the die, x. This is another general feature (as long as the two measures are independent of each other), though not as obvious as for the mean.

It's hard to compare a single roll with the sum of two rolls, but we can compare a single roll with the *average* of two rolls. In this case, they have the *same* mean (we just divide all numbers by 2, so 7 becomes $7/2 = 3.5$). How do the variances and standard deviations compare? The variance of a single roll was $35/12$. The variance of the average of two rolls is similar to the variance of the sum, but all the numbers get halved and then squared, so in fact multiplied by $1/4$, giving $35/24$: half that of one roll. This means that the standard deviation went down as well, but by a factor of $\sqrt{2}$.

If we average three rolls of the die, the variance will go down by a factor of 3 compared to one roll (so would be $35/36$), while the standard deviation would go

---

[21] Pretend one die is red and one is blue. Then the six red values occur with equal probability as do the six blue values, so in total we get $36 = 6 \times 6$ possible values, each with equal probability. A sum of 4 can occur from (Red $\underline{3}$, Blue $\underline{1}$), (Red $\underline{2}$, Blue $\underline{2}$), or (Red $\underline{1}$, Blue $\underline{3}$), so there is a 3 in 36 chance of rolling a 4. The other values are computed similarly.

down by a factor of $\sqrt{3}$. Likewise, after N rolls, the standard deviation becomes very small. This means that the more rolls you average, the more the probabilities will concentrate around the mean of 3.5 and die off rapidly as you move away: so if you roll 1,000 dice and take the average value, it is unlikely to be 3 – though this is perfectly likely for just six rolls, as we have seen.

To illustrate this narrowing better, let's focus on a flip of the coin, which we can think of as a "roll" of a two-sided die. We can think of tails as being $\underline{0}$ and heads being $\underline{1}$. The probabilities are $\left\{\underline{0} : \frac{1}{2}, \underline{1} : \frac{1}{2}\right\}$. The mean is one-half, since $\frac{1}{2} \cdot 0 + \frac{1}{2} \cdot 1 = \frac{1}{2}$. The variance is therefore $\frac{1}{2}\left(0 - \frac{1}{2}\right)^2 + \frac{1}{2}\left(1 - \frac{1}{2}\right)^2 = \frac{1}{4}$. So the standard deviation is $\sqrt{\frac{1}{4}} = \frac{1}{2}$.

With two "rolls" we can take the sum, but as we have seen, it is more fruitful to take the average to compare with a single roll. If you toss a bunch of coins with tails equal to 0 and heads equal to 1, then the average roll is the same as the fraction of coins showing heads, an intuitive concept. We would expect the values to be distributed about the mean of one-half. Repeat: we will flip coins and record their average, then imagine doing this process many times to find the distribution of batch averages. That's a lot to get your head around; maybe it's easier to think of a bunch of college courses, each with a different class average on the final exam. You could ask how those averages are distributed.

Back to our two flips of a coin: if we compute the *average* of two rolls, the probabilities are $\underline{0} : \frac{1}{4}$ (this would be tails-tails), $\frac{1}{2} : \frac{1}{2}$ (one head and one tail), $\underline{1} : \frac{1}{4}$ (both heads). The mean is again $\frac{1}{2}$ and the variance is $\frac{1}{4} \cdot (0 - \frac{1}{2})^2 + \frac{1}{2} \cdot 0^2 + \frac{1}{4} \cdot (1 - \frac{1}{2})^2 = 1/8$, and as anticipated, it went down by a factor of 2, from $1/4$ to $1/8$. The standard deviation has gone down by a factor of $\sqrt{2}$. The figure below plots the distributions for averages of various numbers of flips of the coin.

You can see from the graph that the values are more closely clumped near the mean of one-half when you toss a hundred coins (the black dots). That's because the standard deviation has shrunk by a factor of $\sqrt{100} = 10$ for the black dots: from $1/2$ to $1/20$.

(A4) | If you choose N items independently from a distribution with mean $\overline{x}$ and standard deviation σ, then the batch averages of these N items will have a distribution with mean $\overline{x}$ and standard deviation $\sigma/\sqrt{N}$.

Remark A4 In the graph, the sum of the values for any number of coins must equal 1, since they represent probabilities. Since more potential values are possible with more flips, this explains why the overall size of the graphs is shrinking as the number of coins increases. However, if you look at a *range* of values near the hump, say, the black dots will be individually lower but there are more of them – so the overall probability of finding a result within the range may or may not decrease (depending on where the range is). It is still true, however, that the lower black dots mean that the probability of getting *exactly* half heads and half tails is actually extremely unlikely when you toss a million coins (not shown), but the sharpening

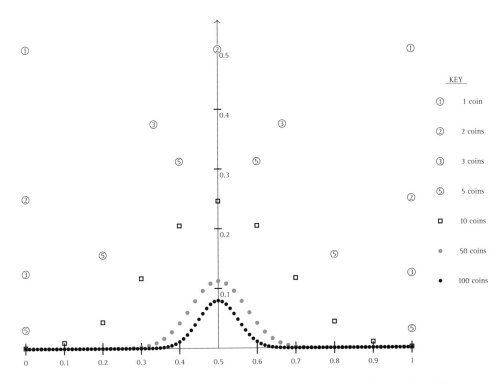

Figure A9 Probabilities for averaging coin flips of 1, 2, 3, 5, 10, 50, and 100 coins. Tails is 0 and Heads is 1, so the *x*-axis is the fraction of coins showing heads.

peaks mean that the *fraction* of coins showing heads will nevertheless be extremely close to 1/2. That's what "probability 1/2" really means. ▲

When the mean is zero and the standard deviation is one, the so-called *normal distribution* is described by the function $\frac{1}{\sqrt{2\pi}}e^{-\frac{x^2}{2}}$, whose graph is the "bell curve" we saw emerging from the black dots in Figure A9. For a general mean of μ (also sometimes denoted $\bar{x}$) and standard deviation σ, the shape is the same but centered around $\bar{x}$ and scaled by σ. The formula is $\frac{1}{\sqrt{2\pi\sigma^2}}e^{-\frac{(x-\mu)^2}{2\sigma^2}}$, pictured as the curve in Figure A10.

If we ask what the probability of measuring something less than x with this distribution of probabilities, the answer is denoted by $\Phi(x)$, the "cumulative distribution function." Although the individual probabilities may decrease toward zero as with the coin tosses, the collective probability of the *range* of values less than x remains finite, equal to $\Phi(x)$. Looking at Figure A10, we see that 0.14% plus 2.14% of the bell curve lies more than two standard deviations below the mean. This means $\Phi(-2) = 0.0014 + 0.0214 = 0.0228$. You can figure out other values from the figure.

To illustrate, return to the picture of the coin tosses. In the figure, there were many little dots, but the graph of the function above is a continuum. We already saw that with more and more dots, the overall height of the probabilities shrinks, as

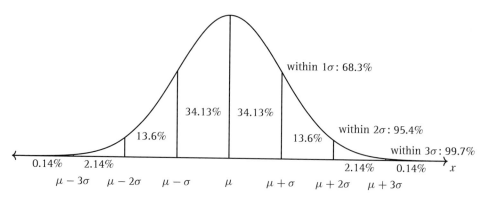

Figure A10 The normal distribution is a familiar "bell curve" centered around a mean of μ with width σ, the standard deviation. The probability of an outcome occurring in some range of x values is equal to the fraction of the curve above those values. For example, the fraction of the curve between one and two standard deviations above the mean is 13.6%. As another example, we see that the probability of falling within one standard deviation of the mean (between $\mu - \sigma$ and $\mu + \sigma$) is 68.3%.

measuring one particular value becomes more and more unlikely. (*Exactly* 133,792 heads out of a million tosses? Unlikely!) But we can talk about the probability of measuring *less* than a particular value, and calculus provides a way[22] of computing this for the normal distribution. The answer is $\Phi(x)$.

What does $\Phi(x)$ look like when x is large and negative? Large and positive?

Think of your response and jot it down.

When x is large and negative, the chance of being less than x is negligible, so Φ is close to 0. When x is very large, it is nearly certain that the values will be less than x, so Φ is close to 1.

We have noticed that the shapes of the gray and black plots seem to be heading toward something smooth and bell shaped. As the number of tosses increases, the shape becomes a *normal distribution*, described by the formula in Figure A10. Remarkably, this is part of a more general phenomenon: if you average *any* random variable (such as coin toss or die value) over a large number of trials, the distribution will concentrate around the mean in a continuous bell-shaped curve called a normal distribution, and this distribution is completely determined by the mean and the standard deviation, or more informally where the bell is and how wide it is. The theorem which asserts this is called the *central limit theorem*. It's amazing, since if you know the mean μ and standard deviation σ for a single event, then if you average N events, the mean will still be μ and the standard deviation will be $\sigma/\sqrt{N}$.

What is $\Phi(0)$ of a normal distribution (with mean value zero)?

Think of your response and jot it down.

The answer is 1/2.

---

[22] In the language of calculus, we have, in fact, $\Phi(x) = \frac{1}{\sqrt{2\pi}} \int_{-\infty}^{x} e^{-u^2/2} du$.

Example A3 Your hometown has a bell tower in its town square that chimes every day at noon. Well, not *exactly* at noon. You decide to measure the time in minutes past (or before) noon, and find that indeed, the mean value is noon (0 minutes past noon), but the distribution follows a normal distribution with standard deviation of 75 s, or 1.25 min. What is the chance that the clocktower will not have rung by 12:05 PM?

To answer this question, we note that 5 min is four standard deviations: in units, 5 min/(1.25 min/std-dev) = 4 std-dev. So we want to know the probability of an event occurring above 4 standard deviations past the mean. That is 1 minus the probability of the event occurring *below* four standard deviations. So the answer is $1 - \Phi(4)$. Cue the calculator! We find $1 - .99997 = 0.00003$, or 3 in a hundred thousand. ▲

Should we have been given a normal distribution, to calculate $\Phi(x)$ we would need to use integral calculus to solve the problem by hand. However, $\Phi(x)$ (also known as the *cumulative distribution function*) can be found on-line or on some calculators using the **normalcdf()** function.

8.6 *p*-Values and the Null Hypothesis

We throw a fair die six times and roll two 3s, no 6s, and every other number once. How do we even know it was a fair die? Maybe it was weighted toward the 3 and against the 6. Maybe there was a manufacturing error and the side that should have been the 6 actually had 3 pips. Can we conclude from our sample of just six throws that the die was unfair?

By now you probably have some intuition that the number of throws was too small to extract a meaningful answer to this question. However, if such a discrepancy – here an average of 3 rather than 3.5 – persisted after a larger number of throws, this would be very unlikely: indeed, the standard deviation would shrink, and 3 would appear to the left of the bell-curve hump. Suppose we had rolled the die a *hundred* times and found an average of 3. The unlikelihood (is that a word?) could be determined if we knew how many standard deviations away from the mean 3 is. Now we saw that the standard deviation for a single coin toss was about 1.7. So batch averages of a hundred coin tosses have a standard deviation of $1.7/\sqrt{100} = 0.17$. The difference between the observed value of the batch average (3) and the mean (3.5) is then 0.5/0.17, or about 2.9 standard deviations away ("2.9σ," we say). In statistics, an event of 3σ or more is considered quite rare.

Remark A5 A few rules of thumb about the normal distribution are useful to know: 68% of measurements will fall within one standard deviation of the mean, 95% within two, and 99.7% within three standard deviations. ▲

What we would like to know is, *with what certainty can we conclude that the die is unfair?*

The way statisticians calculate the answer is to say that we have the *null hypothesis* that the die was fair. We compute the probability of finding a result at least as skewed as ours assuming the null hypothesis, and if this probability is low enough we conclude that the die is unfair.

And how do we find that probability? We just did! We determine how many σs the observation is away from the norm and use the function Φ to compute the probability of finding a deviation that large *or larger*. In the present case, $\Phi(-2.9) \approx 0.00187$, giving this large a chance of recording 2.9 standard deviations or more below the mean. And the same for recording 2.9 standard deviations or more *above* the mean (to find only if the die is weighted to lower values, omit this step) – so a deviation this large or larger would occur with probability 0.00374. We can conclude with 99.6% confidence that the die was not fair.

Often we want to set the confidence level in advance – say, 99%. Then we look up which value of x has $\Phi(x) = 0.005$ and find $x \approx -2.58$, and this means there is a 0.01 chance of measuring a quantity more than 2.58 standard deviations from the mean (above or below). In the case of a hundred throws of the die, 2.58 standard deviations is equal to $2.58 \cdot 0.17 \approx 0.44$, so we could conclude with 99% certainty that any hundred-die average occurring outside of the interval from 3.06 to 3.94 was from an unfair die.

It is up to us to determine the measure of confidence with which we want to state our conclusions.

The tricky part comes when we do not know the exact probabilities or expected distribution of the results. A great deal of statistical theory is set up to deal with potential discrepancies between our sample and the population, as well as all the imperfect knowledge that we have about the quantities we measure. Nevertheless, so that we may gain some familiarity with the quantiative methods of statistics, we will work with examples where some perfect knowledge is assumed.

Remark A6 Your sample is important.

All over the news, media people are making claims based on data collected. Statistics helps you to decide whether there is any basis for the claims. Consider: (1) You check everyone in your family: 0% of them are bald. Is there no baldness in the world? (2) You query all 762 of your friends; 100% of them use Facebook. Is it possible you performed your query with a Facebook post?

These silly examples demonstrate the main pitfalls in analyzing data: the sample may be too small (your family may have escaped baldness by chance, just as you *might* have tossed tails four out of four times), or the sample may not be representative (if you do your query on Facebook, then you are pre-selecting respondents from among Facebook users).

The lessons are clear, but the dangers can be far more subtle. For example, *any* Internet polling will pre-select for Internet users, and while this group includes most Americans, there are plenty without the resources *or the time* to be online, or perhaps the inclination to participate in your survey. Certainly, you will only be able to collect data from among survey participators. If you go offline and collect face-to-face data at a shopping mall, you'll only be sampling shoppers.

And so on. It is an enormous challenge to pollsters and statisticians to ensure they have a *representative sample* of the population about which they want to draw conclusions.

Even if you do find a representative sample, the sample size ("n") may not be large enough to use for meaningful deductions. Anything below 30 is considered "small n" and can be a red flag for drawing strong statistical conclusions. ▲

8.7 Scatterplots and Correlation

"The bigger they are, the harder they fall."

Is it true? To find out, we'd have to collect data. For each one (full disclosure: I don't know what "they" are), we'd find out how big they are and how hard they fell. We can put the results into a scatterplot. If we polled 100 of them, we'd have 100 points to plot. The result might look like Figure A11 or Figure A12.

Sometimes the naked eye can tell that there is some relation, as in Figure A12, but different eyes can interpret data differently, so it would be good to have an objective, quantitative measure of the relationship between the values on one axis and the other. We want a number – called the *correlation* – which is zero if the values are unrelated and one if they are related by a linear function with positive slope. If they are related by a linear function with negative slope we want a correlation of -1. So Figure A11 would give a correlation very near zero, while Figure A12 would have correlation a positive number between zero and one.

To understand the technical definition of correlation, we would like to assume that the x-values and the y-values are chosen from a normal distribution with mean zero and standard deviation one – so if we plotted a very large sample of them it would look a bell curve. If the mean is different we can always shift the values to

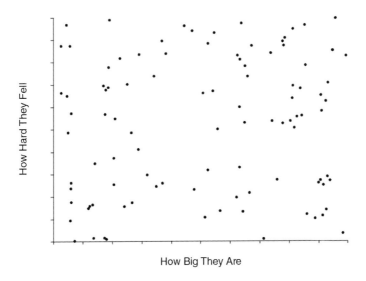

Figure A11 Data Set 1. Is it true that the bigger they are, the harder they fall?

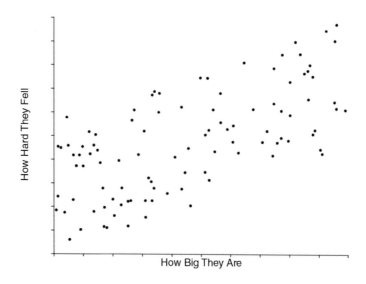

Figure A12 Data Set 2. Is it true that the bigger they are, the harder they fall?

zero, as we did with the test scores at the start of Appendix 8.3. And if the standard deviation σ were not equal to 1, we could divide all values by σ. Now if the y values were a linear function of x then they would have to be $y = mx + b$, but since the mean of the x values and the mean of the y values are both zero, this means b must be zero – so $y = mx$. Now since the x values have standard deviation of 1 and the y values are a multiple m of the x values, the y values must have standard deviation $|m|$, so therefore $m = \pm 1$ if the standard deviation of the y-values is equal to 1. Summarizing, we may assume that both x and y values have norm zero and standard deviation of one. Then perfect correlation means $y = \pm x$.

Now note that standard deviation of 1 means also a variance of 1, which for norm zero means an average value of x^2 equal to 1, or $\langle x^2 \rangle = 1$. Now if $y = x$, this means $\langle xy \rangle = 1$, and if $y = -x$, this means $\langle xy \rangle = \langle x(-x) \rangle = -1$. If the two are unrelated, or independent, then it turns out that the average of the product is the product of their averages, and $\langle xy \rangle = 0$. In fact, the correlation is defined to be simply $\langle xy \rangle$. Had we not rescaled the variables ahead of time, we'd have the unappetizing definition of the correlation as $\frac{\langle (x - \bar{x})(y - \bar{y}) \rangle}{\sigma_x \sigma_y}$.

Example A4 We try a simple (and silly) test of the hypothesis "absence makes the heart grow fonder" by querying people to find out how long it has been since they saw their sweetie and how fond they are feeling. Then we can plot each datum as a point (absence, fondness) and see if there is any correlation. Suppose we do this and find the following data set:

$$\{(2,3),(5,3),(3,6),(4,7),(1,2),(6,5),(10,8),(9,5),(7,7),(2,5),(5,3),(6,6)\},$$

where, say, absence is measured in days and fondness is measured in "heartstrings." Let's leave aside for the moment the fact that this sample is too small to give reliable

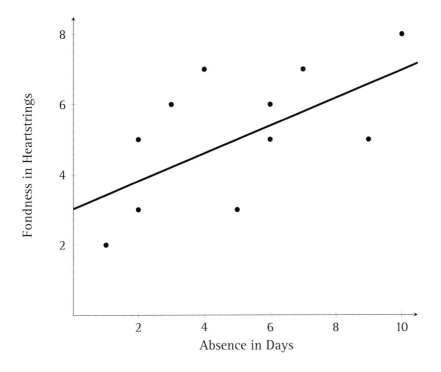

Figure A13 The correlation is about 0.58. The line of least squares fit is shown.

data. If we average the x values we get 5 and if we average the y values we also get 5, so we could use the more complicated formula or we could first shift all values down by 5, giving

$$\{(-3, -2), (0, -2), (-2, 1), (-1, 2), (-4, -3), (1, 0),$$
$$(5, 3), (4, 0), (2, 2), (-3, 0), (0, -2), (1, 1)\}$$

(either method amounts to the same thing). The standard deviations are the square roots of the variances, so averaging the squares of the x values, we get 43/6, and for y we get 10/3, so $\sigma_x = \sqrt{43/6}$ and $\sigma_y = \sqrt{10/3}$, or $\sigma_x\sigma_y = \sqrt{430/18} \approx 4.89$. The formula then tells us that we find the correlation by multiplying the (shifted) x and y values, then taking the average of these numbers and dividing by 4.89. The products of absence and fondness are $\{6, 0, -2, -2, 12, 0, 15, 0, 4, 0, 0, 1\}$ giving an average of $34/12 = 17/6 \approx 2.83$, and a correlation of 0.58.

The purpose of this exercise was to review the algebraic steps needed to calculate correlations. We should compare with what our eyes say when they see the graph.

A correlation between 0.8 and 1 (or negative correlation between -0.8 and -1) is considered to be strong. The number 0.58 is not a very strong correlation. If our data set were larger, we might conclude that there is just a slight relationship between absence and fondness.   ▲

Correlation versus Causation

Suppose you ask a hundred professional tennis players whether they eat eggplants, and you find a strong correlation between the two. Can you conclude that eating eggplants improves your tennis?

The classic mistake of the beginning statistician, and a frequent mistake of news reports, is to confuse correlation with causation.

In our example, perhaps the causation works the other way around, and playing lots of tennis produces a craving for eggplants.

In fact, you can make no conclusions about any causal link. Perhaps the eggplant growers association has a contract with the professional tennis players association requiring members to commit to eating eggplants in order to be eligible for a special end-of-the-year bonus, and this has changed the pros' behavior. Or maybe an eccentric tech billionaire has funded a nationwide program that gives all eggplant eaters free membership at every tennis club in the country, and a lifetime supply of rackets and balls.

This simple example shows the error of confusing causation and correlation in stark relief, but in life it may not be so obvious. The news media presents us with correlations between people in small towns and obscure illnesses, with cellphone use and brain cancer, with chocolates and this or that, with coffee drinking and everything, and either jumps to a conclusion or willfully tempts us to do so.

Correlation is not causation!

8.8 Regression: Whose Line Is It Anyway?

If you are hypothesizing a linear relationship between two variables like absence and fondness, you want to propose an actual equation of a line. This is called a *linear regression*. The most common way to fit a line to your data set is to perform a *least squares* fit. That is, you take the line which gives the smallest value for the average *square* of the vertical distance from the line to the point. For example, the line $y = 5 + \frac{1}{2}x$ passes through the point $(2,6)$, so the data point $(2,3)$ in Example A4 has a vertical distance of 3 (the positive difference of the y values), whose square is 9. Averaging all these quantities will give some number. We want to choose the values of m and b in a line $y = mx + b$ so that that number is as small as possible.

We will not review the formula, except to say that being a quadratic function of m and b, finding the minimum is not much harder than finding the lowest point on a parabola.

Graph it! Anscombe's Quartet

Beware. The mean, standard deviation, correlation and line of least squares fit don't tell the whole story. Anscobe's quartet refers to four data sets, each consisting of eleven points (x,y). For each data set, the mean of x, the variance of x, the mean of y, the variance of y, the correlation between x and y, and the line of best fit are all the same. But the graphs are wildly different. Anscombe introduced these to demonstrate the importance of graphing data.[23]

x_1	y_1	x_2	y_2	x_3	y_3	x_4	y_4
10.0	8.04	10.0	9.14	10.0	7.46	8.0	6.58
8.0	6.95	8.0	8.14	8.0	6.77	8.0	5.76
13.0	7.58	13.0	8.74	13.0	12.74	8.0	7.71
9.0	8.81	9.0	8.77	9.0	7.11	8.0	8.84
11.0	8.33	11.0	9.26	11.0	7.81	8.0	8.47
14.0	9.96	14.0	8.10	14.0	8.84	8.0	7.04
6.0	7.24	6.0	6.13	6.0	6.08	8.0	5.25
4.0	4.26	4.0	3.10	4.0	5.39	19.0	12.50
12.0	10.84	12.0	9.13	12.0	8.15	8.0	5.56
7.0	4.82	7.0	7.26	7.0	6.42	8.0	7.91
5.0	5.68	5.0	4.74	5.0	5.73	8.0	6.89

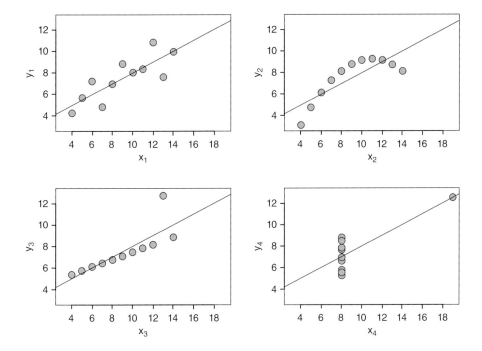

The third data set looks almost perfectly linear, but for that outlier point. Some data analysis is designed to reject outliers, typically after making an objective definition of what an outlier is.[24]

---

[23] F. J. Anscombe, *Graphs in statistical analysis*, The American Statistician **27** (1973) 17–21. Image from Wikipedia: https://en.wikipedia.org/wiki/Anscombe's_quartet.

[24] One such definition takes the value F of the first quartile and the value T of the third quartile, then defines the "inner quartile range" $\Delta = T - F$ to be the difference, and rejects any data point as an outlier if it is 1.5 inner quartile ranges below the first quartile ($< F - 1.5\Delta$) or above the third quartile ($> T + 1.5\Delta$).

Exercises

1 Which of the distributions pictured below has a mean of 5 and a standard deviation of 2?

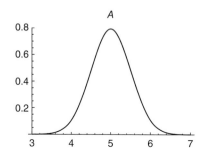

A

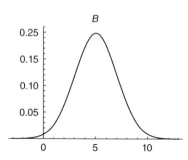

B

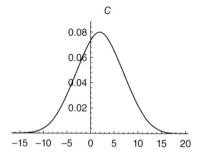

C

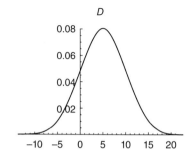

D

2 Calculate the mean, median, variance, and standard deviation of this data set of pulse rates after moderate exercise, measured in beats per minute:

$\{101, 93, 111, 120, 107, 98, 113, 121, 111, 93, 131, 127, 109, 117, 96, 100\}$.

3 The participants in the fictional study of Question 2 were asked to climb some stairs before their heartbeats were measured. The full data set consisted of pairs (heartbeat, number of steps climbed) and is listed here:

$\{(101,30), (93,24), (111,46), (120,49), (107,30), (98,30), (113,35), (121,40)$
$(111,33), (93,25), (131,50), (127,32), (109,40), (117,43), (96,20), (100,25)\}$.

What is the correlation between heartbeat and number of steps climbed? Assuming that our data set represented more samples (for example, each data point might be repeated five times), would you conclude that there is a meaningful link between stairs climbed and heartbeat?

4 Suppose you have a normal distribution with a mean of 6 and a standard deviation of 1.5. Find the number of standard deviations from the mean of the following values:

a 9

b 4.5

c 8.25

d 6.84

5 You toss a coin 10,000 times and get heads 5,120 times. Can you say with
   95% certainty that the coin is weighted toward heads? With 99% certainty?

6 A teacher gives the same test yearly. In the past, the mean score has been 80
   and the standard deviation 12. However, this year the average was 84 with a
   standard deviation of 10. There are 36 students in the class. Allowing $\alpha = 0.01$,
   conduct a null hypothesis test to conclude whether this year's deviation in test
   scores occurred due to chance or other external factors.

9 Estimation

What's the mass of the Earth?

Many people feel overwhelmed by such a question – How could I possibly know?
Where would I even begin? – and sometimes feel panicked and stressed if pressed
to come up with a response.

This kind of question – How many bricks in the Empire State Building? How long
would it take a snail to cross the US? – sometimes called a "Fermi question" after
the great physicist, is a problem designed to test our ability to make back-of-the-
envelope calculations using a few assumptions and approximations. Let's slowly,
calmly, work our way through the mass of the Earth.

What do we need to know to get the mass of the Earth? If we knew its size and
its density, that would be enough. That is, if I know the volume, and I know the
mass per unit volume, then I can multiply the two numbers and get the total mass.
Let's work on those.

Size. How big is the Earth? (No Internet searches!) Maybe you know it's 3,000
miles across the US continent. So how many US's around the Earth? seven? Something between five and ten? So maybe the circumference is between 15,000 and
30,000 miles? Thinking more, maybe you know you cross three time zones travers-
ing the US – so perhaps a time zone is about 1,000 miles, and you cross 24 of
them going around the world. So we've got another estimate of the circumference
as about 24,000 miles, and this does lie between 15,000 and 30,000 miles. So we're
not contradicting ourselves, so far.

Now this circumference is supposed to be pi times the diameter. If we think of pi
as being around 3, this gives a diameter estimate of something between 5,000 and
10,000 miles, with our further estimate of 8. This gives a radius between 2,500 and
5,000 miles, with a shout-out to 4,000.

At this point, we can use the low answer or the high, or just take the 4,000
number. After all, it's just guesswork. Now, what's the volume of a sphere if we know
the radius? Maybe you forget that formula. Well, what if it was just a box? Then
the length of a side would be the diameter, between 5,000 and 10,000 miles, giving
a volume between $5,000^3 = 125,000,000,000$ and $10,000^3 = 1,000,000,000,000$
cubic miles. Hmm ... the sphere would actually fit into that box, so we probably

overestimated. Let's take 500,000,000,000 cubic miles as our working guess. Details won't matter. We're looking for a ballpark.

Density. The Earth does not have a constant density: in some places it's water, in some places it's rock, in some places its iron. Well, there's water on the surface, but not inside, and the inside makes up most of it. In any case, we'll have to guess on what some average density might be. Something between rock and iron. Without looking it up (that would be no fun!), what should we do? Who knows the density of anything, anyway? To know the density of *anything* we would need to know both its mass and its volume. Can you think of something? A book? A can of soda? A laptop? Butter? Each of us will have a different point of contact. I'll take a stick of butter as my reference. I buy four sticks of butter – a pound – and can estimate the block as something like 3 inches by 3 inches by 5 inches. So I have a rough guess that a pound of butter is about 45 cubic inches. A kilogram is a bit more than two pounds, so let's guess that a kilogram of butter is about 100 cubic inches. That will make calculations easy enough. Unfortunately, the Earth is not made of butter. How much heavier than butter is the average piece of the Earth? You probably have a sense that a block of stone is a few times heavier than a block of ice (which is close to the density of water), while a cast iron skillet is somewhat heavier than that still. So let's *guess* that we're looking at the possibility that the average density of the Earth is between 5 and 10 times the density of butter. Where does that leave us? Butter was 1 kg per 100 cubic inches, or 0.01 kg/in^3, so we're guessing between 0.05 and 0.1 kg/in^3.

We're close, but we have a units issue: inches versus miles – 12 inches in a foot, we all know that, and 3 feet in a yard. I know a mile is four times around a running track and once around is 400 somethings, meters or yards. Where does that leave me? A mile is $12 \times 3 \times 4 \times$ (400-plus) inches, or $144 \times$ (400-plus), about what, 60,000 inches? So what's a cubic mile? We have

$$(1 \text{ mi})^3 = (1 \text{ mi} \times 60000 \text{ in/mi})^3 = 216 \times 10^{12} \text{ in}^3,$$

or since 216 is about $200 = 2 \times 10^2$, we can write this as 2×10^{14} in^3. This is a huge number – too hard to guess without starting somewhere more basic.

From above, we have that the volume of the Earth is about 5×10^{11} mi^3, and this gives $5 \times 10^{11} \times 2 \times 10^{14} = 10 \times 10^{25} = 10^{26}$ in^3. A density of 0.1 kg/in^3 then gives a mass of 10^{25} kg. Our lower density guess of 0.05 gives half that, or 5×10^{24} kg. Averaging these, we get about 7.5×10^{24} kg. If you look up the "actual" mass of the Earth as calculated by scientists, you find it is 6×10^{24} kg. Our guess was not bad! Certainly within the ballpark.[25]

It's hard to teach this kind of estimation as a precise skill, but evidently it combines basic skill with weights and measures and geometry as well as units manipulation and scientific notation.

---

[25] I didn't doctor the results, though to be honest I did cube 5,000 wrong at first. (I left out six zeroes the first time I did it.)

Exercises

Try to do each exercise *without* using any resources or looking up any facts!

1 Which quanitity is greater, the number of people on the Earth or the number of water molecules in a gram of water?

   Hint: The mass of one water molecule is approximately 3×10^{-23} grams.
   a Number of people in the world
   b Molecules of water
   c They are roughly equal

2 How many grains of sand are there on the beaches of the world?

3 If you had an ordinary book whose thickness was equal to the distance between the Earth and the Moon, how many pages would it have?

4 How many people are typing the letter Q this very second?

5 How many hairs do you have on your head? Can you argue why there must be two people in Miami who share the same number of follicles? Is this the same same as arguing that there must be someone else with the same number of hairs as you have?

Skills Assessments

This section includes initial and final skills assessments. The initial assessment is for diagnostic purposes, while the final assessment is to take measure of which skills were mastered during the course. These only cover the mathematical skills in the appendices, not the in-depth reasoning of the chapters and projects.

Initial Skills Assessment

Before

The point of this assessment is to help you identify areas of strength and areas that need work heading into this course. This test does not "count." It is purely for your benefit, so you are strongly discouraged from simply guessing an answer or from using a calculator when told not to.

The problems should take no longer than a few minutes each. If a problem takes more than 5 minutes, then even if you can ultimately solve it, you will need to develop a higher comfort level in that subject. Therefore, unless instructed otherwise, don't spend more than 90 minutes on this test.

After

For most of the questions, the topic aligns closely with a section in the Appendices. A key appears below. After you complete the assessment and it has been marked, make a note of which sections posed no difficulty and which you might need to learn anew or review.

Problem(s)	Section(s)	Problem(s)	Section(s)
1	1	15	6, 7
2	1, 2	16	6.1
3	2	17, 18	6.2
4	2.4	19	8.1
5	3.4	20, 21	6
6, 7, 8,	3.5	22	6.5
9, 10	5	23, 24, 25	3.5
11	5.1,5.2	26	7
12	6	27	8
13, 14	6.2		

1 In this question, exact calculations should not be necessary, as a rough guess of the answer is all that is required. For each of the following, circle the correct response.

a 22×0.3 is closest to which of the following numbers?

    A) 1/100   B) 1/10   C) 1   D) 10   E) 100

b $597 \div 2.7$ is closest to which of the following numbers?

    A) 1/100   B) 1/10   C) 1   D) 10   E) 100

c $5280 \div 3600$ is closest to which of the following numbers?

    A) 1/100   B) 1/10   C) 1   D) 10   E) 100

d $.3141592653589 \times 31.41592653589$ is closest to which of the following numbers?

    A) 1/100   B) 1/10   C) 1   D) 10   E) 100

e $(34 + 56) \div 7890$ is closest to which of the following numbers?

    A) 1/100   B) 1/10   C) 1   D) 10   E) 100

2 For each of the following pairs, circle the greater of the two quantities. An exact comparison may not be necessary.

a	10^5	10^{-17}
b	6.023×10^{23}	6023000000
c	123.456	123456×10^{-2}
d	$(525 - 52.5) \times (0.525 + 0.552)$	5×10^3
e	$(5^3)^4$	5^7
f	$(10^3)^5$	10^{3^5}
g	2^3	3^2
h	$\log(10^6)$	10^5
i	$\log(20)$	$\ln(20)$
j	$\log(10^6 \cdot 10^5)$	$\log(10^6) \cdot \log(10^5)$
k	$\ln(4^{5^6})$	$\ln((4^5)^6)$

3 Evaluate the following quantities without using a calculator.

a $2^3 \cdot (4 + 5^2)$

b $\frac{3}{7} + \frac{4}{3}$

c 1.2×0.34

d $(2 - 3)(4 - 5)$

e $(1^2 + 3 \times 4^{5-6})(7 - 8)$

f $\frac{(6^3 \cdot 5^4)^2}{(3^3 \cdot 5^3)^3}$

4 Calculate the following percentages with or without a calculator.

  a 20% of 300

  b 18% of 65

  c 0.12% of 136

  d 3% of 2%

  e 2% of 10% of 50%

  f 375% of 17.39

5 Solve each of the following equations for x.

  a $3x - 4 = 7x + 16$

  b $(x + 2)(x - 3) = x^2 - 2x + 17$

  c $\frac{x+2}{3x-2} = \frac{x+3}{3x+6}$

  d $x^2 - 13x - 40 = 0$

6   a Write an equation relating the number of miles Mary has run (M) to the number of miles Jabari has run (J) if you know that thrice Mary's miles are 70 more than twice Jabari's.

    b Now suppose you also know that Mary has run four times the distance that Jabari has. Write a second equation relating M and J, then solve the pair to determine the miles that Mary and Jabari have run. Label your answer clearly.

7 Nancy's hourly wage is twice Barney's, but Barney works 50% more hours than Nancy. What is the ratio of Nancy's income to Barney's?

8 Molly is 2 years younger than Henry. Six years ago, Molly was two-thirds Henry's age at that time. Let H be Henry's age (now) and let M be Molly's age, both in years. Use the two pieces of information you know to write two equations relating H and M. Solve the equations to determine how old Henry and Molly are.

9 A large egg has about 335 kilojoules (kJ) of energy. A Calorie is about 4.2 kJ. About how many Calories in an egg?

10 It so happens that a speed of 88 feet per second is exactly the same as 60 miles per hour. Knowing only this and the fact that an hour has 3,600 seconds, determine exactly how many feet in a mile.

11 Which of the following is different from the others?

  a 300 million kilometers

  b 3×10^{11} meters

  c 300 trillion centimeters

  d 300000000 kilometers

12 The equations below describe several different animal populations over a period of time. In the equations, P stands for the size of the population and t stands for the time in years.

$$\text{a} \quad P = 1000 - 50t$$

$$\text{b} \quad P = 8000(0.95)^t$$

$$\text{c} \quad P = 1000 + 70t$$

$$\text{d} \quad P = 5000 + 2000\sin(2\pi t)$$

Now for each of the following, write the equation or equations which answer the question.

a This population increases by the same number of animals each year.

b This population decreases by 5% each year.

c This population rises and falls over the course of the year.

d At the time $t = 0$, these populations are the same.

13 Write the equation of the line passing through $(0,32)$ and $(100,212)$. This is the conversion equation for degrees Celsius to degrees Fahrenheit. What is the slope of this line, and what does that say about a Celsius degree?

14 What is the slope of a line passing through the points $(3,7)$ and $(5,-1)$? At what point does it cross the y-axis? What is the equation of this line? At what point does your line cross the x-axis? Given this equation, how can you quickly tell whether or not it will intersect the line $y = x + 1$? Find the point of intersection of these two lines, or state that none exists.

15 I will leave for work within an hour, but I would like to wait for the newspaper to be delivered. If I wait a time T in hours between 0 and 1 to leave, the probability that I will get to work on time is $1 - T$. The probability that I will receive the newspaper before I go is T. The chance that I will be able to bring the paper to work and get there on time is the product of these two quantitites.

a What is the probability of getting the newspaper if I wait a time S in minutes, where $0 \le S \le 60$?

b Write down the probability of getting to work on time with the newspaper as a function of T.

c If you graph this function, what kind of shape will you get?

d What is the line of symmetry for the graph of this function?

e What is the maximum value of this function?

f When should I leave for work if I want to maximize my chance of getting there on time with a paper?

g If I leave at that time, what will my chances be?

h If I will definitely get fired if I'm late, what should I do?

16 The following sketch indicates a company's stock value as a function of time t in months past the new year (so Jan 1 is $t = 0$, Feb 1 is $t = 1$, etc.).

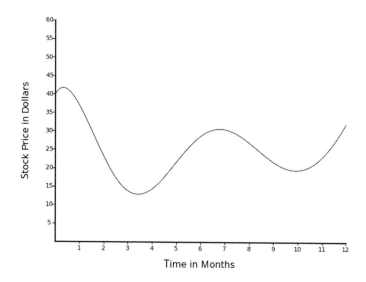

a At approximately what date did the stock reach its highest price?

b About what was the lowest value of the stock?

c On the first of June ($t = 5$), was the value of the stock increasing or decreasing?

d About what dates saw the stock value little changed or unchanged?

e When was the stock's value plummeting most precipitously?

17 Graph the line $y = 7 - 3x$ below. Be sure to label axes, include some hash marks, and label relevant points.

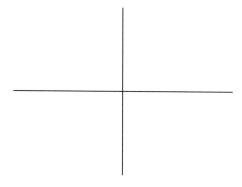

18 George is saving up for a rakish tophat. He starts out his summer with $30 under his mattress and saves $45 each week from his summer job in pest control. Write an equation representing his savings and draw a graph.

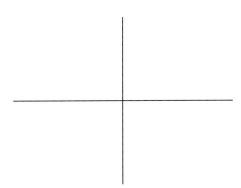

19 What is the average of the whole numbers from 0 to 20?

20 The number of pixels on smartphone displays has been growing steadily over the past few years. Let N be the number of years since the year 2000. Suppose $H(N) = 2000 + 100N$ represents the number of pixels in the horizontal direction on a particular brand of phone, while $V(N) = 1000 + 200N$ represents the number of vertical pixels. By how many total pixels did the resolution of the display increase from 2005 to 2006?

21 The number of ants in my kitchen started out as seven and doubles every three days. Which of the following represents the population N of ants as a function of time t in days? (Circle the correct answer.)

$$\text{a)} \quad N(t) = 2 \cdot 7^{3t} \qquad \text{b)} \quad N(t) = 7 + 2 \cdot 3t$$
$$\text{c)} \quad N(t) = 7 + 2^{3t} \qquad \text{d)} \quad N(t) = 7 \cdot 2^{t/3}$$

22 Solve the following equation for T:

$$e^{-0.07T} = \frac{1}{2}.$$

23 An airline sells 210 tickets for a flight served by an aircraft with capacity 200. Ten percent of ticket holders do not show up for the flight. How many seats will be available for stand-by passengers?

24 A public company has sales of $1.1 billion. Thirty percent of this is profit. Of the profits, $300,000,000 is returned to shareholders as dividend, with the remaining profits invested into research and development ("R&D"). How much will go into R&D?

25 A $12' \times 16'$ rug is cut into $2' \times 2'$ squares, and there are no scraps remaining. How many squares are made?

26 What is the probability of rolling a (total of) 5 with two dice?

27 A class receives grades of 8, 10, 6, 9, 9 on a quiz.
    a What is the mean?
    b What is the median?
    c What is the variance?
    d What is the standard deviation?

Final Skills Assessment

The questions in this final assessment are generally more in-depth than those of the initial assessment, requiring a more comprehensive synthesis of the skills from the appendix. You might allot yourself an hour to complete them.

1 You put $300 in the bank at 4% APR compounded quarterly (four times a year). How much money will you have in the bank after 6 months? Write an expression representing how much money you will have after 5 years.

2 I have 2 liters (2.00 kg) of sugar water, at 3.0% sugar by mass. Sugar is $C_{12}H_{22}O_{11}$, molar mass 342 g/mol. Write an expression for the number of sugar molecules there are in the solution.

Hint: Avogadro's number, 6.02×10^{23}, is the number of molecules per mole.

3 Newspapers appear sporadically upon my doorstep. Each day, there is a one-half chance that no newspaper will appear, a one-third chance that one newspaper will appear, and a one-sixth chance that two newspapers will appear. I am going away for two months (60 days). How many newspapers can I expect to be piled at my doorstep upon my return?

4 Calculate the mean, variance, and standard deviation of the data set $\{1,2,3,4,5\}$. Would these quantities change if the numbers one through five represented values of a fair, five-sided die, and mean, variance, and standard deviation were taken in the probabilistic sense? If, instead of rolling a single die, you were to roll a dozen and record the average number, would the distribution of averages have a smaller, larger, or the same standard deviation as for a single roll?

Hint: recall that the variance is the average squared-separation from the mean.

Glossary

absolute value. If x is a number – positive or negative – the absolute value $|x|$ measures its "size," so is always greater than or equal to zero. If $x \geq 0$, then $|x| = x$, but if $x < 0$, $|x| = -x$. Note $|x| \geq 0$ for any x, and $|x| = 0$ if and only if $x = 0$.

annual percentage rate (APR). An annual rate on which loan interest calculations are based. For example, if the APR were 12%, then interest compounded monthly would be one-twelfth of that, or 1%. Daily interest would be 0.12/365. Because accrued interest will generate further interest in later compoundings, the APR does not actually represent the interest accumulated in one year, unless interest is only compounded once. *See* annual percentage yield (APY).

annual percentage yield (APY). A measure of the interest on a loan that is earned in one year, including interest-upon-interest as a result of multiple compoundings. For example, a loan with an APR of 8.0% that is compounded monthly will grow by a factor of $(1 + 0.08/12)$ each month, or by $(1 + 0.08/12)^{12} = 1.083$ in a year. Since the loan grows by 8.3% in a year, we say the annual percentage yield is 8.3%. More generally, if the APR is r (written as a fraction, not a percentage), and interest is compounted N times in a year, then the APY is $(1 + r/N)^N - 1$.

area. A measure of the "size" of a two-dimensional shape. The area of a rectangle is the product of its length times its width. The units of area are therefore length-squared, e.g., cm^2.

associativity. The property that expresses the notion that it doesn't matter how you group pairs of things to combine more than two. In arithmetic, it is the property of addition that $(a+b)+c = a+(b+c)$ for all numbers a,b,c. This allows you to write $a+b+c$ unambiguously. In multiplication, it is the property $(ab)c = a(bc)$. We take this for granted, but *cooking*, for example, is not associative! When making bread, you need to combine water and yeast and flour. If you do ((water+yeast)+flour), you'll be fine, but (water + (yeast + flour)) won't work.

average. Given a set of N numbers (data set), the average – also called the mean – is the sum of them divided by N. If we write the numbers as $x_1, x_2, \ldots, x_N$, then the average is denoted $\bar{x}$ or $\langle x \rangle$, so $\bar{x} = \frac{1}{N}(x_1 + \cdots + x_N)$.

circumference. The length around a circle. If the radius is R, the circumference is $2\pi R$.

coefficient. The number in front of one of several expressions, e.g., as in a polynomial. For example, if $f(x) = 3x^7 + 6x^3 + 2$, the coefficient of the x^3 term is 6.

commutativity. The property that expresses the notion that it doesn't matter which order you combine things. For example, addition and multiplication are commutative because $a + b = b + a$ and $a \times b = b \times a$, but subtraction is not commutative because $a - b \neq b - a$.

confounder. In statistics, a variable that correlates (either positively or negatively) with a variable being examined. For example, suppose you measure the effect of sunglasses on how random people's beauty is perceived. If it turns out that beautiful people are more likely to wear sunglasses in the first place, then the person's beauty will be a confounding factor.

conversion factor. A number that you multiply to convert a measurement from one unit to another. For example, since 1 in = 2.54 cm, dividing both sides by "1 in" gives $1 = 2.54$ cm/in. This is a conversion factor for going from inches to centimeters, since, for example, 12 in = 12 in $\times$ 1 = 12 in $\times$ 2.54 cm/in = 30.48 cm. Note that the "inches" cancel.

correlation. Given a set of pairs of numbers $(x_1, y_1), \ldots, (x_N, y_N)$, the correlation between them measures the degree to which the y values depend on the x values through a linear relationship (or vice versa). We denote the correlation $\rho_{x,y}$ and define it by $\rho_{x,y} = \frac{\langle (x - \bar{x})(y - \bar{y}) \rangle}{\sigma_x \sigma_y}$, where the overlines indicate mean value and σ indicates the standard deviation.

cost–benefit. A scheme for analyzing the economic value of a decision. One tallies the benefits (measured or anticipated) and subtracts the cost of implementation and operation. For example, if a company wants to expand its sales force, the cost–benefit analysis will weigh the anticipated gain in revenues from sales against the costs of hiring, paying, and managing the new employees.

cumulative distribution function. The function, $\Phi(x)$, describing the probability of measuring a value less than x for a random variable obeying a normal distribution with given mean and standard deviation. In this book, we only consider a mean of zero and standard deviation of 1, so we take x to be measured in "number of standard deviations above the mean," or Z-score.

dimensional analysis. The process of estimating a quantity's magnitude by determining what units it must have, then forming a combination of other quantities in the problem that have the same units.

distributivity. The arithmetic property that $a(b + c) = ab + ac$ for all numbers a, b, c. This property can be used to prove FOIL: $(p + q)(r + s) = (p + q)r + (p + q)s$, then use distributivity twice more (after writing the last two terms in the opposite order using commutativity) to write as $pr + qr + ps + qs$, which can be rearranged using commutativity to obtain the FOIL ordering, $pr + ps + qr + qs$.

expected value. Given a set of values $x_1, \ldots, x_N$ occurring with probabilities $p_1, \ldots, p_N$, the expected value $\langle x \rangle$ is the weighted sum $p_1 x_1 + \cdots p_N x_N$. Note that when all values are equally likely, the expected value agrees with the mean, justifying the use of the same notation.

exponential, exponential growth, exponential decay. Exponential growth or decay refers to the behavior of members of a population (such as a population of cancer cells, mold spores, dollars of a stock price, immigrants, fruit flies, electric charge, radioactive isotopes) over time. Exponential growth means that in a fixed period of time, the population increases by a fixed factor greater than 1. Decay means that the factor is less than 1. If we call time t and the population P, then if the function $P(t)$ is exponential, it can be written as $P(t) = ab^t$. Exponential growth means $b > 1$, decay means $b < 1$. We can also write $P(t) = P_0 e^{kt}$, where $P_0 = a$ and $e^k = b$ or $k = \ln(b)$, in which case, growth or decay occurs according to whether $k > 0$ or $k < 0$, respectively.

factor. In multiplication, one of a product of terms – so x is a factor of xy and $a + b$ is a factor of $(a + b)(a - c)$.

factorial. $N! = N \cdot (N - 1) \cdot (N - 2) \cdots 2 \cdot 1$. For example, $5! = 120$. The quantity $N!$ counts the number of ways of ordering N objects. For example, $3! = 6$ is the number of "words" you can form from 3 letters: ABC, ACB, BCA, BAC, CAB, CBA. Note: $0! = 1$.

function. A rule for assigning a number to an element from an input set. Typically, inputs are also numbers. We denote the value a function f assigns to a number x as $f(x)$. This notation can be used to define the function as well, for example, $f(x) = x^3$ defines the function that takes a numerical input and outputs its cube. For this case, e.g., $f(4) = 64$. The input/output process is sometimes denoted $x \mapsto f(x)$, here $4 \mapsto 64$.

graph. A visual representation of a function or data set. The graph of a function $f(x)$ is a set of points (x,y) satisfying $y = f(x)$.

half-life. The time it takes for an exponential decay function, such as the measure of radioactivity over time, to decay to half its value. If the function is $f(t) = Ce^{-at}$, the half-life T satisfies $f(t+T) = \frac{1}{2}f(t)$. This yields $e^{-aT} = \frac{1}{2}$, or $T = \ln(2)/a$. The bigger the decay rate a, the shorter the half-life.

independent event. An event or occurrence whose probability does not depend on prior events. For example, if you roll one die and then roll it again, the second roll is an independent event from the first one.

integer. A whole number or the negative of a whole number: $\ldots -2, -1, 0, 1, 2, \ldots$. An integer-valued variable is often (though not exclusively) denoted by n or m, as distinct from x or y, which may indicate real numbers. The set of all integers is usually denoted as $\mathbb{Z}$.

interest. The fee paid to a lender for the privilege of borrowing, typically a percentage of the value borrowed. (A bank pays interest to account holders.)

irrational number. A number that cannot be written as a fraction, i.e., cannot be written as m/n, where m and n are integers. The decimal expansion of an irrational number goes on forever without a repeating pattern.

linear. A relationship between input and output of some process is linear if the output is a linear function of the input, meaning any plot of input values and output values would lie on a straight line. It also means that if you increase the input by a fixed amount, the output will always increase by a fixed multiple of that amount. For example, the number of tires is a linear function of the number of bikes (i.e., twice).

linear function. A polynomial with no term involving a power greater than 1. A function $f(x)$ of the form $f(x) = ax + b$ for some numbers a and b.

logarithm. The inverse function to the exponential function. That is, $\log 10^x = x$; so, for example, $\log 1{,}000 = 3$. The natural logarithm is defined by $\ln e^x = x$.

mean. *See* average, expected value.

median. Given a set of numbers (data set), the median is the value for which half the data are above and half are below. If there is no single such value – as for the data set $\{1,3,5,7,8,8\}$ – then the median is the average of the two closest values. Here 5 and 7 are the closest, so the median is 6.

molarity. The concentration of a compound in solution, expressed in units of molars, or moles per liter, and denoted M. For example, consider 56 grams of table salt (sodium chloride: NaCl) dissolved in 1 liter of water. Since the atomic mass of Na is 11 and that of Cl is 17, 56 grams represents two moles ($2 \times 28 = 56$). This solution would have molarity of about 2 mol/L = 2M. We write $[NaCl] = 2M$.

mole. A number of molecules ("Avogadro's number"), about equal to the number of hydrogen atoms in 1 gram, approximately 6.022×10^{23}; more precisely, the number of carbon-12 atoms in 12 grams. The mass, measured in grams, of 1 mole of a molecule is approximately equal to the sum of the atomic masses of the constituents, measured in atomic mass units. For example, 1 mole of water (H_2O) has a mass of about $2 \times 1 + 8 = 10$ grams.

molecule. A group of atoms linked by chemical bonds. For example, a molecule of water (H_2O) consists of two hydrogen atoms and an oxygen atom.

normal distribution. A bell-curve-shaped function describing the probability of measuring a quantity, characterized by the mean and standard deviation. The central limit theorem says that large batch averages of a quantity are described by a normal distribution. In formulas, the normal distribution of a quantity (or "random variable") x with mean $\bar{x}$ and standard deviation σ is $\frac{1}{\sqrt{2\pi\sigma^2}} e^{-\frac{((x-\bar{x})^2}{2\sigma^2}}$.

null hypothesis. The assumption that the treatment in a trial had no effect. If, for a known distribution, this is found to happen with a probability (p-value) less than some chosen threshold or tolerance, then the null hypothesis will be rejected in favor of the alternative hypothesis: that the treatment had an effect.

opportunity cost. When you choose to spend your time and money doing one thing, you simultaneously choose *not* to spend your time and money on other things. The opportunity cost is the measure of the most valuable thing you have not done with your time and money. If you had to forgo viewing a video of a cute kitten to read this definition, then I hope that the value of understanding opportunity cost is greater than the value in pleasure you would have felt watching Kitty struggle with a ball of yarn.

order of magnitude. A power of ten that approximates a number; 890 has order of magnitude 3, since it is close to 10^3. Two numbers can be said to have the same order of magnitude if they differ by a factor roughly less than 5, such as 13 and 46; otherwise, they differ in order of magnitude by the order of magnitude of their ratio. The number 1,623 is about two orders of magnitude higher than the number 13, as their ratio is about 125, which has order of magnitude 2.

outlier. A data point that doesn't follow the pattern of the rest of the data, such as one that is more than three standard deviations away from the mean or from a linear regression.

***p*-value.** The probability that collected data support the "null hypothesis," which is that the data occur just by chance. For example, the data may show a decline in measured fish population in the world's fisheries. It could be a coincidence of sudden deaths (the null hypothesis), or there may be something causing the fish to die (the alternative hypothesis). The statistician will want to set a *significance level*, such as 1% or 5%. If the *p*-value is below the significance level, the null hypothesis will be rejected in favor of the alternative hypothesis.

parabola. A "bowl-shaped" curve in the plane that focuses lines from infinity, such as the graph of a quadratic function $ax^2 + bx + c$. (The surface of rotation is a paraboloid, a shape used in reflecting telescopes or headlights for its focusing properties.) A parabola can be defined as the set of points equidistant from a chosen focus and a given line (*directrix*). For the curve $y = ax^2 + bx + c$, first define the *discriminant* $\Delta = b^2 - 4ac$; then the focus lies at $(-\frac{b}{2a}, \frac{1-\Delta}{4a})$, and the directrix is the line $y = -\frac{1+\Delta}{4a}$. A parabola can be formed by slicing a cone with any plane parallel to one of the rays forming the cone, or as the trajectory of a particle subject to a constant gravitational force.

percent. Literally, "out of a hundred," so, for example, "twelve percent" means "twelve out of a hundred," i.e., $12/100$ or 0.12, denoted 12%. For example, 3% of 250 is $0.03 \times 250 = 7.5$. Changes and errors are often measured in percentages, as in "the Dow Jones Industrial Average fell 2% today" or "your guess of 8,670 jelly beans was off by just two percent: there were 8,500."

permutation. An ordered arrangement of a set of distinct objects. For example, $ADCB$ is a permutation of the letters $\{A, B, C, D\}$. If there are N objects, then there are $N!$ ("N factorial") permutations. Sometimes "permutation" refers to ordered arrangements including just k of the N objects, in which case, there are $N!/(N-k)!$ such permutations. For example, when $k = 2$ in our example, there are $4!/2! = 4 \times 3$ such arrangements: $AB, AC, AD, BA, BC, BD, CA, CB, CD, DA, DB, DC$.

polynomial. A function of a specific form: for functions of one variable, a polynomial is a sum of (nonnegative) powers of the variable multiplied by a number (coefficient), such as $17x^7 + 4x^3 + 2x + 6$. The highest power appearing is called the "degree." For polynomials in more than one variable, replace "powers of the variable" by "products of powers of the variables" in the definition: for example, $13xy^2z - 3yz^3 + xyz$ is a polynomial in three variables.

proportionality. A linear relationship between quantities so that one is a constant multiple of the other. For example, sales receipts are proportional to the number of theatergoers: if each ticket cost $13, then T theatergoers will generate sales of $S = 13T$. Here 13 is the "constant of proportionality."

Pythagorean Theorem. An equation expressing the relationship between the lengths of sides of a right triangle. If a and b are the leg lengths and c is the length of the hypotenuse, then $c^2 = a^2 + b^2$. For instance, if a point (x,y) is a distance 13 from the origin $(0,0)$ and you know $y = 12$, then since $13^2 = x^2 + 12^2$, we conclude $x = \sqrt{169 - 144} = 5$.

quadratic formula. An expression for the roots of a quadratic function: $ax^2 + bx + c = 0$ is solved by $x = \frac{1}{2a}\left(-b \pm \sqrt{b^2 - 4ac}\right)$. For example, the roots of $x^2 - 4x + 3$ are $\frac{1}{2}\left(4 \pm \sqrt{16 - 12}\right)$, or 1 and 3.

quadratic function. A polynomial of degree two in one variable, such as $ax^2 + bx + c$.

qualitative. Involving concepts but not quantities. A qualitative understanding of gravity holds that massive objects attract each other, explaining why the Earth stays in orbit around the Sun.

quantitative. Involving numbers. A quantitative understanding of gravity is that the attractive force between two massive objects is proportional to the product of their masses and inversely proportional to the square of the distance between them.

regression. An approximation of a data set of pairs of values by a specified form of curve. A *linear regression* is an approximate linear relationship between the values, often formed by the line of "least squares fit." For example, if the data are a set of points $(x_1,y_1),(x_2,y_2),\ldots,(x_n,y_n)$, then the line of least squares fit is the line $y = mx + b$ determined by choosing m and b so as to minimize the sum $(y_1 - (mx_1 + b))^2 + \cdots + (y_n - (mx_n + b))^2$.

risk–reward. An assessment of the likely outcome of a decision by evaluating the potential risks/costs involved against the possible rewards/benefits/gains. The assessment may involve a probabilistic model for potential outcomes.

root. A zero of a function, i.e., an input value for which the function returns zero. For example, the roots of the function $f(x) = x^3 - x$ are $x = -1,0,1$, as can be seen by factoring: $x^3 - x = x(x^2 - 1) = x(x + 1)(x - 1)$.

scatterplot. Given a data set of pairs $(x_1,y_1),\ldots,(x_N,y_N)$, the scatterplot of them is the set of all points $(x_1,y_1),\ldots,(x_N,y_N)$ drawn in the plane. It is used as a visual aid in representing a correlation, or lack thereof.

scientific notation. A method of expressing a number, particularly a very large or small one, which highlights its order of magnitude. For example, scientific notation for 0.00000465 would be 4.65×10^{-6}. The number that multiplies the power of 10 must be at least 1 and less than 10.

selection bias. The failure to ensure a representative sample for data collection. For example, a survey about e-commerce that is conducted online will be biased by selecting only Internet users.

significant digits, significant figures. The measure of precision for values/ quantities/measurements used in analysis. Counting significant digits is simplest in scientific notation: the number 4.65×10^{-6} is given to three significant digits, namely, the 4, 6, and 5 in 4.65. The number 0.000032 has two significant digits, since it is 3.2×10^{-5}. To indicate the same quantity with one more significant digits, you might write 0.0000320 or 3.20×10^{-5}.

slope. A measure of the rate of increase of a linear function; the steepness of a line. A function has slope m if, for each unit increase of the input, the output increases by m. For example, the line $y = mx + b$ has slope m since $m(x + 1) + b$ is precisely m units larger than $mx + b$, for any value of x. The line, therefore, has constant slope or rate of change. The "instantaneous rate of change" of a function $f(x)$ at a value x_0 is represented as the slope of the line tangent to the curve $y = f(x)$ at the point $(x_0, f(x_0))$.

standard deviation. Given a set of N numbers $x_1, \ldots, x_N$ such as a data set, the standard deviation σ_x is a measure of the spread from the mean. It is defined by $\sigma_x = \sqrt{\langle (x - \bar{x})^2 \rangle}$, in other words, the square root of the variance. For example, the standard deviation of the data set 2,4,5,6,8 is 2. *See* **variance**.

variable. An unknown or unprescribed quantity denoted with a letter or symbol, as in "let the variable x represent the number of acres of corn grown in Iowa." Variables can also represent input values for functions, as in "the operation of addition can be represented as a function of two variables: $f(x, y) = x + y$."

variance. Given a set of N numbers $x_1, \ldots, x_N$ such as a data set, the variance is a measure of the spread from the mean. It is the average of the squared distance from the mean, defined by $\text{var}(x) = \langle (x - \bar{x})^2 \rangle$. For example, the data set 2,4,5,6,8 has mean 5, so the variance is the average of 9,1,0,1,9, which is 4.

volume. A measure of the "size" of a three-dimensional object or region. The volume of a box is the product of its length, width, and height. The units of volume are therefore length-cubed, e.g., cm^3.

Z-score. For a measured value x, its Z-score is the number of standard deviations above the mean, so $Z = (x - \langle x \rangle)/\sigma$. For example, if the mean is 12 and the standard deviation is 2, a measured value of 9 has a Z-score of $(9 - 12)/2 = -1.5$.

Index